TOPOLOGY:
A BEGINNER'S GUIDE

TOPOLOGY: A BEGINNER'S GUIDE

Rohit Gupta

Oxford Book Company
Jaipur ▪ New Delhi

Published By

Oxford Book Company

Regd. Off. "Bony Residency", Gate No. 2
Opp. Tilak Public School, Vishweswariya Nagar,
Gopalpura Road, Jaipur - 302018, India
Ph: 0141 - 2761280, Telefax: 2761381
email - oxfordbookjpr@gmail.com
website - oxfordbookcompany.com

Delhi Office: 102, 1st Floor, Satyam House,
4327/3, Ansari Road, Darya Ganj,
New Delhi - 110002, India
Ph: 011-45652440

Distributed By
KEYA INTERNATIONAL PRESS LLP
E-mail: keyainternationalpressllp@gmail.com

First published in India, 2026

ISBN : 978-93-5524-295-2

Typeset by Shivangi Computers, Jaipur

Printed and bound in India by Thomson Press Ltd.,
New Delhi

Contents

Preface

The mathematical study of qualities that are maintained when objects are stretched, twisted, and deformed is known as topology. However, tears are not permitted. A sphere is topologically identical to an ellipsoid, and a circle is topologically equivalent to an ellipse, into which it can be stretched. In a similar vein, the set of all possible positions for a clock's hour hand is topologically equivalent to a circle (a closed curve in one dimension without any intersections that can be embedded in two dimensions); the set of all possible positions for the hour and minute hands combined is topologically equivalent to a torus's surface (a two-dimensional surface that can be embedded in three dimensions); and the set of all possible positions for the hour, minute, and second hands combined is topologically equivalent to a three-dimensional object.

It is possible to utilize topology to abstract an object's intrinsic connectedness while disregarding its specific shape. For instance, the connection of several topologically different surfaces is depicted in the above diagrams. Corners labeled with the same letter correspond to the same location in these diagrams because parallel edges depicted in solid join one another with the orientation indicated by arrows, while dashed lines indicate edges that stay free. Topological spaces are frequently used to formally define the "objects" of topology. Two things are considered to be homeomorphic if they share the same topological characteristics.

Poincaré developed homotopy, a measure of an object's topology, around 1900 (Collins 2004). Specifically, if one may be continually deformed into the other, then two mathematical objects are said to be homotopic.

There are three types of topology: low-dimensional topology, differential topology, and algebraic topology. Point-set topology is the low-level language of topology, which isn't truly regarded as a distinct "branch" of topology. It is sometimes possible to utilize topology to solve algebraic problems, even though algebraic topology mostly uses algebra to analyze topological difficulties. For instance, it is easy to demonstrate that any subgroup of a free group is also a free group in algebraic topology. The fundamental approach now used in algebraic topology is to study spaces through algebraic invariants by, for instance, transferring spaces to groups with a lot of controllable structure while maintaining the homeomorphism of spaces relation. This makes it possible to transform claims about topological spaces into claims about groups, which are frequently simpler to prove. There are two main methods for doing this: homology and cohomology groups, and basic groups, or more generally, homotopy theory. Although the fundamental groups are frequently nonabelian and can be challenging to work with, they provide us with vital information about the structure of a topological space. There is a finite presentation of the basic group of a (finite) simplicial complex. In contrast, homology and cohomology groups are abelian and finitely formed in many significant circumstances. Finitely generated abelian groups are especially simple to work with and are fully categorized.

This book on topology covers both algebraic and generic topology in detail. There are numerous instances and figures presented. It will be very beneficial to anyone who needs a thorough, basic introduction to algebraic and general topology and its applications.

— ***Rohit Gupta***

Chapter 1

Topology Basics

Algebraic Geometry

Since classical algebraic geometry allows to talk about polynomial equations, but not inequalities, the *algebraic standard n-simplex* is commonly defined as the subset of affine n+1-dimensional space, where all coordinates sum up to 1 (thus leaving out the inequality part). The algebraic description of this set is

$$\Delta^n := \{x \in \mathbb{A}^{n+1} \mid \sum_{i=1}^{n+1} x_i - 1 = 0\},$$

which equals the scheme-theoretic description $\Delta_n(R) = Spec(R[\Delta^n])$ with

$$R[\Delta^n] := R[x_1, \ldots, x_{n+1}] / (\sum x_i - 1)$$

the ring of regular functions on the algebraic n-simplex (for any ring R). By using the same definitions as for the classical n-simplex, the n-simplices for different dimensions n $R[\Delta^n]$ assemble into one simplicial object, while the rings assemble into one cosimplicial object $R[\Delta^\bullet]$ (in the category of schemes resp. rings, since the face and degeneracy maps are all polynomial). The algebraic n-simplices are used in higher K-Theory and in the definition of higher Chow groups.

Applications

Simplices are used in plotting quantities that sum to 1, such as proportions of subpopulations, as in a ternary plot.

In industrial statistics, simplices arise in problem formulation and in algorithmic solution. In the design of bread, the producer must combine yeast, flour, water, sugar, etc. In such mixtures, only the relative proportions of ingredients matters: For an optimal bread mixture, if the flour is doubled then the yeast should be doubled. Such mixture problem are often formulated with normalized constraints, so that the nonnegative components sum to one, in which case the feasible region forms a simplex. The quality of the bread mixtures can be estimated using response surface methodology, and then a local maximum can be computed using a nonlinear programming method, such as sequential quadratic programming. In operations research, linear programming problems can be solved by the simplex algorithm of George Dantzig. In geometric design and computer graphics, many methods first perform simplicial triangulations of the domain and then fit interpolating polynomials to each simplex.

Topological Space

Topological spaces are mathematical structures that allow the formal definition of concepts such as convergence, connectedness, and continuity. They appear in virtually every branch of modern mathematics and are a central unifying notion. The branch of mathematics that studies topological spaces in their own right is called topolog

Definition

The utility of the notion of a topology is shown by the fact that there are several equivalent definitions of this structure. The most commonly used, and the most elegant, is that in terms of *open sets*, but the most intuitive is that in terms of *neighbourhoods* and so we give this first.

Neighbourhoods Definition

Let X be a set; the elements of X are usually called *points*, though they can be any mathematical objects. We allow X to be empty. Let N be a function assigning to each x in a non-empty set N(x) of subsets of . The elements of N(x) will be called *neighbourhoods* of with respect to N (or, simply, *neighbourhoods of* x). The function N is called a neighbourhood topology if the axioms below are satisfied; and then X with N is called a topological space. A topological space in which the *points* are functions is called a function space.

1. If N is a neighbourhood of x, then $x \in N$.
2. If N is a subset of X containing a neighbourhood of x, then is a neighbourhood of x.
3. The intersection of two neighbourhoods of is again a neighbourhood of .
4. Any neighbourhood $N$of x contains a neighbourhood M of x such that N is a neighbourhood of each point of M.

The first three axioms for neighbourhoods have a clear meaning. The fourth axiom has a very important use in the structure of the theory, that of linking together the neighbourhoods of different points of X.

A standard example of such a system of neighbourhoods is for the real line R, where a subset N of R is defined to be a *neighbourhood* of a real number if there is an open interval containing x and contained in N.

Open Sets Definition

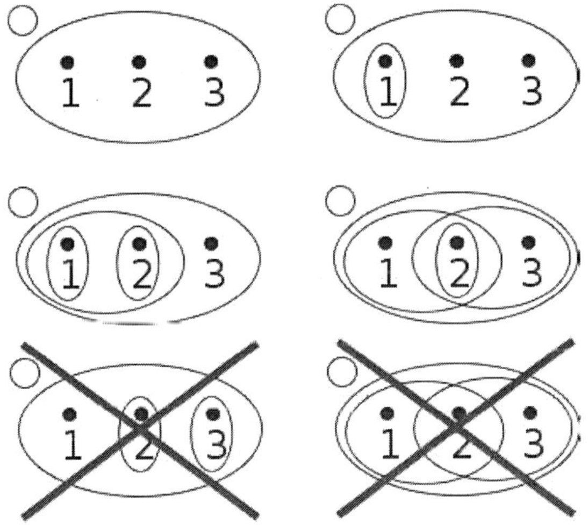

Figure: *Four examples and two non-examples of topologies on the three-point set {1,2,3}. The bottom-left example is not a topology because the union of {2} and {3} [i.e. {2,3}] is missing; the bottom-right example is not a topology because the intersection of {1,2} and {2,3} [i.e. {2}], is missing.*

Given such a structure, we can define a subset U of X to be open if U is a neighbourhood of all points in U. It is a remarkable fact that the open sets then satisfy the elegant axioms given below, and that, given these axioms, we can recover the neighbourhoods satisfying the above axioms by defining N to be a neighbourhood of x if N contains an open set U such that $x \in U$.

A *topological space* is then a set X together with a collection of subsets of X, called open sets and satisfying the following axioms:

1. The empty set and X itself are open.
2. Any union of open sets is open.
3. Any finite intersection of open sets is open.

The collection of open sets is then also called a topology on X, or, if more precision is needed, an open set topology. The sets in are called the open sets, and their complements in X are called closed sets. A subset of X may be neither closed nor open, either closed or open, or both. A set that is both closed and open is called a clopen set.

Characterizations of the Category of Topological Spaces

In mathematics, a topological space is usually defined in terms of open sets. However, there are many equivalent characterizations of the category of topological spaces. Each of these definitions provides a new way of thinking about topological concepts, and many of these have led to further lines of enquiry and generalisation.

Definitions

Formally, each of the following definitions defines a concrete category, and every pair of these categories can be shown to be *concretely isomorphic*. This means that for every pair of categories defined below, there is an isomorphism of categories, for which corresponding objects have the same underlying set and corresponding morphisms are identical as set functions.

To actually establish the concrete isomorphisms is more tedious than illuminating. The simplest approach is probably to construct pairs of inverse concrete isomorphisms between each category and the category of topological spaces Top. This would involve the following:

1. Defining inverse object functions, checking that they are inverse, and checking that corresponding objects have the same underlying set.
2. Checking that a set function is "continuous" (i.e., a morphism) in the given category if and only if it is continuous (a morphism) in Top.

Definition Via Open Sets

Objects: all topological spaces, i.e., all pairs (X,T) of set X together with a collection T of subsets of X satisfying:

1. The empty set and X are in T.
2. The union of any collection of sets in T is also in T.
3. The intersection of any pair of sets in T is also in T.

The sets in T are the open sets.

Morphisms: all ordinary continuous functions, i.e. all functions such that the inverse image of every open set is open.

Comments: This is the ordinary category of topological spaces.

Definition Via Closed Sets

Objects: all pairs (X,T) of set X together with a collection T of subsets of X satisfying:

1. The empty set and X are in T.
2. The intersection of any collection of sets in T is also in T.
3. The union of any pair of sets in T is also in T.

The sets in T are the closed sets.

Morphisms: all functions such that the inverse image of every closed set is closed.

Comments: This is the category that results by replacing each lattice of open sets in a topological space by its order-theoretic dual of closed sets, the lattice of complements of open sets. The relation between the two definitions is given by De Morgan's laws.

Definition Via Closure Operators

Objects: all pairs (X,cl) of set X together with a closure operator cl : $P(X) \to P(X)$ satisfying the Kuratowski closure axioms:

1. $A \subseteq \text{cl}(A)$ (Extensivity)
2. $\text{cl}(\text{cl}(A)) = \text{cl}(A)$ (Idempotence)
3. $\text{cl}(A \cup B) = \text{cl}(A) \cup \text{cl}(B)$ (Preservation of binary unions)
4. $\text{cl}(\varnothing) = \varnothing$ (Preservation of nullary unions)

Morphisms: all *closure-preserving functions*, i.e., all functions f between two closure spaces

$$f : (X, \text{cl}) \to (X', \text{cl}')$$

such that for all subsets A of X

$$f(\text{cl}(A)) \subset \text{cl}'(f(A))$$

Comments: The Kuratowski closure axioms abstract the properties of the closure operator on a topological space, which assigns to each subset its topological closure. This topological closure operator has been generalized in category theory.

Definition Via Interior Operators

Objects: all pairs (X,int) of set X together with an interior operator int : $P(X) \to P(X)$ satisfying the following dualisation of the Kuratowski closure axioms:

1. $A \supseteq \text{int}(A)$
2. $\text{int}(\text{int}(A)) = \text{int}(A)$ (Idempotence)
3. $\text{int}(A \cap B) = \text{int}(A) \cap \text{int}(B)$ (Preservation of binary intersections)
4. $\text{int}(X) = X$ (Preservation of nullary intersections)

Morphisms: all *interior-preserving functions*, i.e., all functions f between two interior spaces

$$f : (X, \text{int}) \to (X', \text{int}')$$

such that for all subsets A of X'

$$f^{-1}(\text{int}'(A)) \subset \text{int}(f^{-1}(A))$$

Comments: The interior operator assigns to each subset its topological interior, in the same way the closure operator assigns to each subset its topological closure.

Definition Via Neighbourhoods

Objects: all pairs (X,N) of set X together with a neighbourhood function $N : X \to F(X)$, where $F(X)$ denotes the set of all filters on X, satisfying for every x in X:

1. If U is in $N(x)$, then x is in U.
2. If U is in $N(x)$, then there exists V in $N(x)$ such that U is in $N(y)$ for all y in V.

Morphisms: all *neighbourhood-preserving functions*, i.e., all functions $f : (X, N) \to (Y, N')$ such that if V is in $N(f(x))$, then there exists U in $N(x)$ such that $f(U)$ is contained in V. This is equivalent to asking that whenever V is in $N(f(x))$, then $f^{-1}(V)$ is in $N(x)$.

Comments: This definition axiomatizes the notion of neighbourhood. We say that U is a neighbourhood of x if U is in $N(x)$. The open sets can be recovered by declaring a set to be open if it is a neighbourhood of each of its points; the final axiom then states that every neighbourhood contains an open set. These axioms (coupled with the Hausdorff condition) can be retraced to Felix Hausdorff's original definition of a topological space in Grundzüge der Mengenlehre.

Definition Via Convergence

The category of topological spaces can also be defined via a convergence relation between filters on X and points of x. This definition demonstrates that convergence of filters can be viewed as a fundamental topological notion.

A topology in the usual sense can be recovered by declaring a set A to be closed if, whenever F is a filter on A, then A contains all points to which F converges.

Similarly, Top can be described via net convergence. As for filters, this definition shows that convergence of nets can be viewed as a fundamental topological notion. A topology in the usual sense can be recovered by declaring a set A to be closed if, whenever (x_α) is a net on A, then A contains all points to which (x_α) converges.

Comparison of Topologies

In topology and related areas of mathematics comparison of topologies refers to the fact that two topological structures on a given set may stand in relation to each other. The set of all possible topologies on a given set forms a partially ordered set. This order relation can be used to compare the different topologies.

Definition

Let (either the open or closed sets does not matter) τ_1 and τ_2 be two topologies on a set X such that τ_1 is contained in τ_2:

$$\tau_1 \subseteq \tau_2 .$$

That is, every element of τ_1 is also an element of τ_2. Then the topology τ_1 is said to be a coarser (weaker or smaller) topology than τ_2, and τ_2 is said to be a finer (stronger or larger) topology than τ_1. If additionally

$$\tau_1 \neq \tau_2$$

we say τ_1 is strictly coarser than τ_2 and τ_2 is strictly finer than τ_1. The binary relation $\subseteq$ defines a partial ordering relation on the set of all possible topologies on X.

Examples

The finest topology on X is the discrete topology. The coarsest topology on X is the trivial topology. In function spaces and spaces of measures there are often a number of possible topologies. All possible polar topologies on a dual pair are finer than the weak topology and coarser than the strong topology.

Lattice of Topologies

The set of all topologies on a set X together with the partial ordering relation $\subseteq$ forms a complete lattice. That is, any collection of topologies on X have a *meet* (or infimum) and a *join* (or supremum). The meet of a collection of topologies is the intersection of those topologies. The join, however, is not generally the union of those topologies (the union of two topologies need not be a topology) but rather the topology generated by the union.

Every complete lattice is also a bounded lattice, which is to say that it has a greatest and least element. In the case of topologies, the greatest element is the discrete topology and the least element is the trivial topology.

Specializations and generalizations

The following spaces and algebras are either more specialized or more general than the topological spaces discussed above.

- Proximity spaces provide a notion of closeness of two sets.
- Metric spaces embody a metric, a precise notion of distance between points.
- Uniform spaces axiomatize ordering the distance between distinct points.
- Cauchy spaces axiomatize the ability to test whether a net is Cauchy. Cauchy spaces provide a general setting for studying completions.
- Convergence spaces capture some of the features of convergence of filters.
- Grothendieck sites are categories with additional data axiomatizing whether a family of arrows covers an object. Sites are a general setting for defining sheaves.

Proximity Space

In topology, a proximity space is an axiomatization of notions of "nearness" that hold set-to-set, as opposed to the better known point-to-set notions that characterize topological spaces.

The concept was described by Frigyes Riesz in 1908 and ignored at the time. It was rediscovered and axiomatized by V. A. Efremoviè in 1934, but not published until 1951. In the interim, in 1940, A. N. Wallace discovered a version of the same concept.

Definition A proximity space (X, Δ) is a set X with a relation Δ between subsets of X satisfying the following properties:

For all subsets A, B and C of X

$$A\delta B \Rightarrow B\delta A$$

$$A\delta B \Rightarrow A \neq \emptyset$$

$$A \cap B \neq \emptyset \Rightarrow A \boldsymbol{\delta} B$$

$$A \boldsymbol{\delta} (B \cup C) \Leftrightarrow (A \boldsymbol{\delta} B \text{or} A \boldsymbol{\delta} C)$$

$$(\forall E, A \boldsymbol{\delta} E \text{or} B \boldsymbol{\delta} (X - E)) \Rightarrow A \boldsymbol{\delta} B$$

Proximity without the first axiom is called quasi-proximity. If $A\ \delta\ B$ we say A is near B or A and B are proximal. We say B is a proximal or Δ-neighbourhood of A, written A « B, if and only if $A\ \delta\ X$-B is false. The main properties of this set neighbourhood relation, listed below, provide an alternative axiomatic characterization of proximity spaces.

For all subsets A, B, C, and D of X,

1. X«X
2. A«$B \Rightarrow A \subseteq B$
3. $A \subseteq B$«$C \subseteq D \Rightarrow A$«$D$
4. (A«B and A«C) $\Rightarrow A$«B) $\cap C$
5. A«$B \Rightarrow X$-B«X"A
6. A«$B \Rightarrow$ "E, A«E«B

A proximity space is called *separated* if $\{x\}\ \delta\ \{y\}$ implies x=y.

A *proximity* or *proximal map* is one that preserves nearness, that is, given f:(X,Δ)$\rightarrow$(X^*,δ^*), if $A\ \delta\ B$ in X, then $f[A]\ \delta^*\ f[B]$ in X^*. Equivalently, a map is proximal if the inverse map preserves proximal neighbourhoodness. In the same notation, this means if C«$^*D$ holds in X^*, then $f^1[C]$«$f^1[D]$ holds in X.

Given a proximity space, one can define a topology by letting $A \rightarrow \{x : \{x\}\ \delta\ A\}$ be a Kuratowski closure operator. If the proximity space is separated, the resulting topology is Hausdorff. Proximity maps will be continuous between the induced topologies. The resulting topology is always completely regular. This can be proven by imitating the usual proofs of Urysohn's lemma, using the last property of proximal neighbourhoods to create the infinite increasing chain used in proving the lemma. Given a compact Hausdorff space, there is a unique proximity whose corresponding topology is the given topology: A is near B if and only if their closures intersect. More generally, proximities classify the compactifications of a completely regular Hausdorff space.

A uniform space X induces a proximity relation by declaring A is near B if and only if $A \times B$ has nonempty intersection with every entourage. Uniformly continuous maps will then be proximally continuous.

Metric Space

In mathematics, a metric space is a set where a notion of distance (called a metric) between elements of the set is defined. The metric space which most closely corresponds to our intuitive understanding of space is the 3-dimensional Euclidean space. In fact, the notion of "metric" is a generalization of the Euclidean metric arising from the four long-known properties of the Euclidean distance. The Euclidean metric defines the distance between two points as the length of the straight line segment connecting them. Non-intuitive metric spaces occur in elliptic geometry and hyperbolic geometry. For example, the hyperboloid model of hyperbolic geometry is used in special relativity for a metric space of velocities. A metric space also induces topological properties like open and closed sets which leads to the study of even more abstract topological spaces.

History

Maurice Fréchet introduced metric spaces in his work *Sur quelques points du calcul fonctionnel*, Rendic. Circ. Mat. Palermo 22 (1906) 1–74.

Examples of Metric Spaces

- Ignoring mathematical details, for any system of roads and terrains the distance between two locations can be defined as the length of the shortest route connecting those locations. To be a metric there shouldn't be any one-way roads. The triangle inequality expresses the fact that detours aren't shortcuts. Many of the examples below can be seen as concrete versions of this general idea.
- The real numbers with the distance function $d(x, y) = | y - x |$ given by the absolute difference, and more generally Euclidean n-space with the Euclidean distance, are complete metric spaces. The rational numbers with the same distance also form a metric space, but are not complete.

- The *positive* real numbers with distance function $d(x, y) = \| y - x \|$ is a complete metric space.
- Any normed vector space is a metric space by defining, see also relation of norms and metrics. (If such a space is complete, we call it a Banach space.) Examples:
  - o The Manhattan norm gives rise to the Manhattan distance, where the distance between any two points, or vectors, is the sum of the differences between corresponding coordinates.
  - o The maximum norm gives rise to the Chebyshev distance or chessboard distance, the minimal number of moves a chess king would take to travel from to x.y
- The British Rail metric (also called the Post Office metric or the SNCF metric) on a normed vector space is given by $d(x, y) = \| x \| + \| y \|$ for distinct points x and y, and $d(x, x) = 0$. More generally $\| . \|$ can be replaced with a f function taking an arbitrary set to non-negative reals and taking the value 0at most once: then the metric is defined on by $d(x, y) = f(x) + f(y)$ for distinct points and x, y and $d(x, x) = 0$ The name alludes to the tendency of railway journeys (or letters) to proceed via London (or Paris) irrespective of their final destination.
- If (M, d) is a metric space and X is a subset of M, then (X, d) becomes a metric space by restricting the domain of d to $X \times X$.
- The discrete metric, where $d(x, y) = 0$ if $x = y$ and $d(x, y) = 1$ otherwise, is a simple but important example, and can be applied to all non-empty sets. This, in particular, shows that for any non-empty set, there is always a metric space associated to it. Using this metric, any point is an open ball, and therefore every subset is open and the space has the discrete topology.
- A finite metric space is a metric space having a finite number of points. Not every finite metric space can be isometrically embedded in a Euclidean space.

- The hyperbolic plane is a metric space. More generally:
- If M is any connected Riemannian manifold, then we can turn M into a metric space by defining the distance of two points as the infimum of the lengths of the paths (continuously differentiable curves) connecting them.
- If x is some set and M is a metric space, then, the set of all bounded functions $f: X \to M$ (i.e. those functions whose image is a bounded subset of M) can be turned into a metric space by defining $d(f,g) = \sup_{x \in X} d(f(x), g(x))$ for any two bounded functions f and g (where sup is supremum. This metric M is called the uniform metric or supremum metric, and If is complete, then this function space is complete as well. If X is also a topological space, then set of all bounded continuous functions from X to M (endowed with the uniform metric), will also be a complete metric if M is.
- If G is an undirected connected graph, then the V set of vertices of G can be turned into a metric space by defining $d(x,y)$ to be the length of the shortest path connecting the vertices X and y. In geometric group theory this is applied to the Cayley graph of a group, yielding the word metric.
- The Levenshtein distance is a measure of the dissimilarity between two strings u and u, defined as the minimal number of character deletions, insertions, or substitutions required to transform u into u. This can be thought of as a special case of the shortest path metric in a graph and is one example of an edit distance.
- Given a metric space (X,d) and an increasing concave function $f:[0,\infty) \to [0,\infty)$ such that if and only $f(x) = 0$ if $x = 0$, then $f°d$ is also a metric on X.
- Given an injective function f from any set A to a metric space (X,d), $d(f(x), f(y))$ defines a metric on A.
- Using T-theory, the tight span of a metric space is also a metric space. The tight span is useful in several types of analysis.

- The set of all m by n matrices over some field is a metric space with respect to the rank distance $d(X,Y) = \operatorname{rank}(Y - X)$.
- The Helly metric is used in game theory.

Bounded and Totally Bounded Spaces

A metric space *M* is called bounded if there exists some number *r*, such that $d(x,y) \leq r$ for all *x* and *y* in *M*. The smallest possible such *r* is called the diameter of *M*. The space *M* is called precompact or totally bounded if for every $r > 0$ there exist finitely many open balls of radius *r* whose union covers *M*. Since the set of the centres of these balls is finite, it has finite diameter, from which it follows (using the triangle inequality) that every totally bounded space is bounded. The converse does not hold, since any infinite set can be given the discrete metric (one of the examples above) under which it is bounded and yet not totally bounded.

Note that in the context of intervals in the space of real numbers and occasionally regions in a Euclidean space R^n a bounded set is referred to as "a finite interval" or "finite region". However boundedness should not in general be confused with "finite", which refers to the number of elements, not to how far the set extends; finiteness implies boundedness, but not conversely.

Compact Spaces

A metric space *M* is compact if every sequence in *M* has a subsequence that converges to a point in *M*. This is known as sequential compactness and, in metric spaces (but not in general topological spaces), is equivalent to the topological notions of countable compactness and compactness defined via open covers.

Examples of compact metric spaces include the closed interval [0,1] with the absolute value metric, all metric spaces with finitely many points, and the Cantor set. Every closed subset of a compact space is itself compact. A metric space is compact iff it is complete and totally bounded. This is known as the Heine–Borel theorem. Note that compactness depends only on the topology, while boundedness depends on the metric.

Lebesgue's number lemma states that for every open cover of a compact metric space M, there exists a "Lebesgue number" δ such that every subset of M of diameter $< \delta$ is contained in some member of the cover. Every compact metric space is second countable, and is a continuous image of the Cantor set. (The latter result is due to Pavel Alexandrov and Urysohn.)

Locally Compact and Proper Spaces

A metric space is said to be locally compact if every point has a compact neighbourhood. Euclidean spaces are locally compact, but infinite-dimensional Banach spaces are not.

A space is proper if every closed ball $\{y : d(x,y) \leq r\}$ is compact. Proper spaces are locally compact, but the converse is not true in general.

Connectedness

A metric space M is connected if the only subsets that are both open and closed are the empty set and M itself.

A metric space M is path connected if for any two points $x, y \in M$ there exists a continuous map $f:[0,1] \to M$ with $f(0) = x$ and $f(1) = y$. Every path connected space is connected, but the converse is not true in general. There are also local versions of these definitions: locally connected spaces and locally path connected spaces.

Simply connected spaces are those that, in a certain sense, do not have "holes".

Separable Spaces

A metric space is separable space if it has a countable dense subset. Typical examples are the real numbers or any Euclidean space. For metric spaces (but not for general topological spaces) separability is equivalent to second countability and also to the Lindelöf property.

Types of Maps Between Metric Spaces

Suppose (M_1,d_1) and (M_2,d_2) are two metric spaces.

Continuous Maps

The map $f:M_1 \to M_2$ is continuous if it has one (and therefore all) of the following equivalent properties:

General topological continuity

For every open set U in M_2, the preimage $f^{-1}(U)$ is open in M_1

This is the general definition of continuity in topology.

Sequential continuity

if (x_n) is a sequence in M_1 that converges to x in M_1, then the sequence $(f(x_n))$ converges to $f(x)$ in M_2.

This is sequential continuity, due to Eduard Heine.

ε-δ definition

for every x in M_1 and every $\varepsilon>0$ there exists $\delta>0$ such that for all y in M_1 we have

$$d_1(x,y)<\delta \Rightarrow d_2(f(x),f(y))<\varepsilon.$$

This uses the (ε, δ)-definition of limit, and is due to Augustin Louis Cauchy.

Moreover, f is continuous if and only if it is continuous on every compact subset of M_1.

The image of every compact set under a continuous function is compact, and the image of every connected set under a continuous function is connected.

Uniformly Continuous Maps

The map $f : M_1 \to M_2$ is uniformly continuous if for every $\varepsilon > 0$ there exists $\delta > 0$ such that

$$d_1(x,y)<\delta \Rightarrow d_2(f(x),f(y))<\varepsilon \quad \text{for all} \quad x,y \in M_1.$$

Every uniformly continuous map $f : M_1 \to M_2$ is continuous. The converse is true if M_1 is compact (Heine–Cantor theorem).

Uniformly continuous maps turn Cauchy sequences in M_1 into Cauchy sequences in M_2. For continuous maps this is generally wrong; for example, a continuous map from the open interval (0,1) *onto* the real line turns some Cauchy sequences into unbounded sequences.

Lipschitz-continuous maps and contractions

Given a number $K > 0$, the map $f : M_1 \to M_2$ is K-Lipschitz continuous if

$$d(f(x),f(y))<d(x,y) \quad \text{for all} \quad x \neq y \in M_1$$

Every Lipschitz-continuous map is uniformly continuous, but the converse is not true in general.

If $K < 1$, then f is called a contraction. Suppose $M_2 = M_1$ and M_1 is complete. If f is a contraction, then f admits a unique fixed point (Banach fixed point theorem).

If M_1 is compact, the condition can be weakened a bit: f admits a unique fixed point if

$$d(f(x), f(y)) < d(x, y) \quad \text{for all} \quad x \neq y \in M_1 .$$

Isometries

The map $f{:}M_1 \to M_2$ is an isometry if

$$d_2(f(x), f(y)) = d_1(x, y) \quad \text{for all} \quad x, y \in M_1$$

Isometries are always injective; the image of a compact or complete set under an isometry is compact or complete, respectively. However, if the isometry is not surjective, then the image of a closed (or open) set need not be closed (or open).

Quasi-isometries

The map $f : M_1 \to M_2$ is a quasi-isometry if there exist constants $A \geq 1$ and $B \geq 0$ such that

$$\frac{1}{A} d_2(f(x), f(y)) - B \leq d_1(x, y) \leq A d_2(f(x), f(y)) + B \text{ for all } x, y \in M_1$$

and a constant $C \geq 0$ such that every point in M_2 has a distance at most C from some point in the image $f(M_1)$.

Note that a quasi-isometry is not required to be continuous. Quasi-isometries compare the "large-scale structure" of metric spaces; they find use in geometric group theory in relation to the word metric.

Notions of Metric Space Equivalence

Given two metric spaces (M_1, d_1) and (M_2, d_2):

- They are called homeomorphic (topologically isomorphic) if there exists a homeomorphism between them (i.e., a bijection continuous in both directions).
- They are called uniformic (uniformly isomorphic) if there exists a uniform isomorphism between them (i.e., a bijection uniformly continuous in both directions).

- They are called isometric if there exists a bijective isometry between them. In this case, the two metric spaces are essentially identical.
- They are called quasi-isometric if there exists a quasi-isometry between them.

Topological properties

Metric spaces are paracompact Hausdorff spaces and hence normal (indeed they are perfectly normal). An important consequence is that every metric space admits partitions of unity and that every continuous real-valued function defined on a closed subset of a metric space can be extended to a continuous map on the whole space (Tietze extension theorem). It is also true that every real-valued Lipschitz-continuous map defined on a subset of a metric space can be extended to a Lipschitz-continuous map on the whole space.

Metric spaces are first countable since one can use balls with rational radius as a neighbourhood base.

The metric topology on a metric space M is the coarsest topology on M relative to which the metric d is a continuous map from the product of M with itself to the non-negative real numbers.

Distance Between Points and Sets; Hausdorff Distance and Gromov Metric

A simple way to construct a function separating a point from a closed set (as required for a completely regular space) is to consider the distance between the point and the set. If (M,d) is a metric space, S is a subset of M and x is a point of M, we define the distance from x to S as

$d(x,S)=\inf\{d(x,s):s\in S\}$ where $\inf$ represents the infimum.

Then $d(x, S) = 0$ if and only if x belongs to the closure of S. Furthermore, we have the following generalization of the triangle inequality:

$$d(x,S)\leq d(x,y)+d(y,S),$$

which in particular shows that the map $x\mapsto d(x,S)$ is continuous.

Given two subsets S and T of M, we define their Hausdorff distance to be

$$d_H(S,T) = \max\{\sup\{d(s,T) : s \in S\}, \sup\{d(t,S) : t \in T\}\}$$

where $\sup$ represents the supremum.

In general, the Hausdorff distance $d_H(S,T)$ can be infinite. Two sets are close to each other in the Hausdorff distance if every element of either set is close to some element of the other set.

The Hausdorff distance d_H turns the set $K(M)$ of all non-empty compact subsets of M into a metric space. One can show that $K(M)$ is complete if M is complete. (A different notion of convergence of compact subsets is given by the Kuratowski convergence.)

One can then define the Gromov–Hausdorff distance between any two metric spaces by considering the minimal Hausdorff distance of isometrically embedded versions of the two spaces. Using this distance, the set of all (isometry classes of) compact metric spaces becomes a metric space in its own right.

Product Metric Spaces

If $(M_1, d_1), \ldots, (M_n, d_n)$ are metric spaces, and N is the Euclidean norm on R^n, then $\left(M_1 \times \ldots \times M_n, N(d_1, \ldots, d_n)\right)$ is a metric space, where the product metric is defined by

$$N(d_1,\ldots,d_n)\big((x_1,\ldots,x_n),(y_1,\ldots,y_n)\big) = N\big(d_1(x_1,y_1),\ldots,d_n(x_n,y_n)\big),$$

and the induced topology agrees with the product topology. By the equivalence of norms in finite dimensions, an equivalent metric is obtained if N is the taxicab norm, a p-norm, the max norm, or any other norm which is non-decreasing as the coordinates of a positive n-tuple increase (yielding the triangle inequality).

Similarly, a countable product of metric spaces can be obtained using the following metric

$$d(x,y) = \sum_{i=1}^{\infty} \frac{1}{2^i} \frac{d_i(x_i,y_i)}{1 + d_i(x_i,y_i)}.$$

An uncountable product of metric spaces need not be metrizable. For example, $\mathbf{R}^{\mathbf{R}}$ is not first-countable and thus isn't metrizable.

Continuity of Distance

It is worth noting that in the case of a single space (M,d), the distance map $d: M \times M \to R^{+}$ (from the definition) is uniformly continuous with respect to any of the above product metrics $N(d,d)$, and in particular is continuous with respect to the product topology of $M \times M$.

Generalizations of Metric Spaces

- Every metric space is a uniform space in a natural manner, and every uniform space is naturally a topological space. Uniform and topological spaces can therefore be regarded as generalizations of metric spaces.
- If we consider the first definition of a metric space given above and relax the second requirement, we arrive at the concepts of a pseudometric space or a dislocated metric space. If we remove the third or forth, we arrive at a quasimetric space, or a semimetric space.
- If the distance function takes values in the extended real number line R∪{+”}, but otherwise satisfies all four conditions, then it is called an *extended metric* and the corresponding space is called an ∞*-metric space*. If the distance function takes values in some (suitable) ordered set (and the triangle inequality is adjusted accordingly), then we arrive at the notion of *generalized ultrametric.*
- Approach spaces are a generalization of metric spaces, based on point-to-set distances, instead of point-to-point distances.
- A continuity space is a generalization of metric spaces and posets, that can be used to unify the notions of metric spaces and domains.

Uniform Space

In the mathematical field of topology, a uniform space is a set with a uniform structure. Uniform spaces are topological spaces with additional structure which is used to define uniform properties such as completeness, uniform continuity and uniform convergence.

The conceptual difference between uniform and topological structures is that in a uniform space, one can formalize certain notions of relative closeness and closeness of points. In other words, ideas like "x is closer to a than y is to b" make sense in uniform spaces. By comparison, in a general topological space, given sets A,B it is meaningful to say that a point x is *arbitrarily close* to A (i.e., in the closure of A), or perhaps that A is a *smaller neighbourhood* of x than B, but notions of closeness of points and relative closeness are not described well by topological structure alone.

Uniform spaces generalize metric spaces and topological groups and therefore underlie most of analysis.

Definition

There are three equivalent definitions for a uniform structure.

Entourage Definition

A uniform space (X, Φ) is a set X equipped with a nonempty family Φ of subsets of the Cartesian product $X \times X$ (Φ is called the uniform structure or uniformity of X and its elements entourages (French: neighbourhoods or *surroundings*)) that satisfies the following axioms:

1. if U is in Φ then U contains the diagonal D = $\{ (x, x) : x - X \}$.
2. if U is in Φ and V is a subset of $X \times X$ which contains U, then V is in Φ
3. if U and V are in F, then $U \cap V$ is in Φ
4. if U is in Φ then there exists V in Φ such that, whenever (x, y) and (y, z) are in V, then (x, z) is in U.
5. if U is in Φ then $U^{-1} = \{ (y, x) : (x, y) \text{ in } U \}$ is also in Φ.

If the last property is omitted we call the space quasiuniform.

One usually writes $U[x]=\{y : (x,y) \in U\}$. On a graph, a typical entourage is drawn as a blob surrounding the "$y=x$" diagonal;

the $U[x]$'s are then the vertical cross-sections. If $(x,y) \in U$, one says that x and y are *U-close*. Similarly, if all pairs of points in a subset A of X are U-close (i.e., if $A \times A$ is contained in U), A is called *U-small*. An entourage U is *symmetric* if $(x,y) \in U$ precisely when $(y,x) \in U$. The first axiom states that each point is U-close to itself for each entourage U. The third axiom guarantees that being "both U-close and V-close" is also a closeness relation in the uniformity. The fourth axiom states that for each entourage U there is an entourage V which is "half as large". Finally, the last axiom states the essentially symmetric property "closeness" with respect to a uniform structure.

A fundamental system of entourages of a uniformity ϑ is any set B of entourages of Φ such that every entourage of T contains a set belonging to B. Thus, by property 2 above, a fundamental systems of entourages B is enough to specify the uniformity Φ unambiguously: Φ is the set of subsets of $X \times X$ that contain a set of B. Every uniform space has a fundamental system of entourages consisting of symmetric entourages. The right intuition about uniformities is provided by the example of metric spaces: if (X,d) is a metric space, the sets

$$U_a = \{(x,y) \in X \times X : d(x,y) \leq a\} \quad \text{where} \quad a > 0$$

form a fundamental system of entourages for the standard uniform structure of X. Then x and y are U_a-close precisely when the distance between x and y is at most a.

A uniformity Φ is *finer* than another uniformity Ø on the same set if $\Phi \supseteq \psi$; in that case ψ is said to be *coarser* than Φ.

Pseudometrics Definition

Uniform spaces may be defined alternatively and equivalently using systems of pseudometrics, an approach which is particularly useful in functional analysis (with pseudometrics provided by seminorms). More precisely, let f: $X \times X \to \mathrm{R}$ be a pseudometric on a set X. The inverse images $U_a = f^{-1}([0,a])$ for $a > 0$ can be shown to form a fundamental system of entourages of a uniformity. The uniformity generated by the U_a is the uniformity defined by the single pseudometric f. Certain authors call spaces the topology of which is defined in terms of pseudometrics *gauge spaces*.

For a *family* (f_i) of pseudometrics on X, the uniform structure defined by the family is the *least upper bound* of the uniform structures defined by the individual pseudometrics f_i. A fundamental system of entourages of this uniformity is provided by the set of *finite* intersections of entourages of the uniformities defined by the individual pseudometrics f_i. If the family of pseudometrics is *finite*, it can be seen that the same uniform structure is defined by a *single* pseudometric, namely the *upper envelope* sup f_i of the family.

Less trivially, it can be shown that a uniform structure that admits a countable fundamental system of entourages (and hence in particular a uniformity defined by a countable family of pseudometrics) can be defined by a single pseudometric. A consequence is that *any* uniform structure can be defined as above by a (possibly uncountable) family of pseudometrics.

Uniform Cover Definition

A uniform space (X,Θ) is a set X equipped with a distinguished family of *uniform covers* Θ from the set of coverings of X, forming a filter when ordered by star refinement. One says cover P is a *star refinement* of cover Q, written P <* Q, if for every $A \in$ P, there is a $U \in$ Q such that if $A \cap B$ `$\in$ ø, $B \in$ P, then $B \subseteq U$. Axiomatically, this reduces to:

1. {X} is a uniform cover.
2. If P <* Q and P is a uniform cover, then Q is also a uniform cover.
3. If P and Q are uniform covers, then there is a uniform cover R that star-refines both P and Q.

Given a point x and a uniform cover P, one can consider the union of the members of P that contain x as a typical neighbourhood of x of "size" P, and this intuitive measure applies uniformly over the space.

Given a uniform space in the entourage sense, define a cover P to be uniform if there is some entourage U such that for each $x \in X$, there is an $A \in$ P such that $U[x] \subseteq A$.

These uniform covers form a uniform space as in the second definition. Conversely, given a uniform space in the uniform cover sense, the supersets of $\Gamma \in \{A \times A : A \in \text{P}\}$, as P ranges over the uniform covers, are the entourages for a uniform space as in the first definition. Moreover, these two transformations are inverses of each other.

Topology of uniform spaces

Every uniform space X becomes a topological space by defining a subset O of X to be open if and only if for every x in O there exists an entourage V such that $V[x]$ is a subset of O. In this topology, the neighbourhood filter of a point x is $\{V[x] : V \in \Phi\}$. This can be proved with a recursive use of the existence of a "half-size" entourage. Compared to a general topological space the existence of the uniform structure makes possible the comparison of sizes of neighbourhoods: $V[x]$ and $V[y]$ are considered to be of the "same size".

The topology defined by a uniform structure is said to be induced by the uniformity.

A uniform structure on a topological space is *compatible* with the topology if the topology defined by the uniform structure coincides with the original topology. In general several different uniform structures can be compatible with a given topology on X.

Uniformizable Spaces

A topological space is called uniformizable if there is a uniform structure compatible with the topology.

Every uniformizable space is a completely regular topological space. Moreover, for a uniformizable space X the following are equivalent:

- X is a Kolmogorov space
- X is a Hausdorff space
- X is a Tychonoff space
- for any compatible uniform structure, the intersection of all entourages is the diagonal $\{(x, x) : x \text{ in } X\}$.

Some authors (e.g. Engelking) add this last condition directly in the definition of a uniformizable space. The topology of a uniformizable space is always a symmetric topology; that is, the space is an R_0-space.

Conversely, each completely regular space is uniformizable. A uniformity compatible with the topology of a completely regular space X can be defined as the coarsest uniformity which makes all continuous real-valued functions on X uniformly continuous.

A fundamental system of entourages for this uniformity is provided by all finite intersections of sets $(f \times f)^{-1}(V)$, where f is a continuous real-valued function on X and V is an entourage of the uniform space R.

This uniformity defines a topology, which is clearly coarser than the original topology of X; that it is also finer than the original topology (hence coincides with it) is a simple consequence of complete regularity: for any $x \in X$ and a neighbourhood V of x, there is a continuous real-valued function f with $f(x)=0$ and equal to 1 in the complement of V.

In particular, a compact Hausdorff space is uniformizable. In fact, for a compact Hausdorff space X the set of all neighbourhoods of the diagonal in $X \times X$ form the *unique* uniformity compatible with the topology.

A Hausdorff uniform space is metrizable if its uniformity can be defined by a *countable* family of pseudometrics. Indeed, as discussed above, such a uniformity can be defined by a *single* pseudometric, which is necessarily a metric if the space is Hausdorff. In particular, if the topology of a vector space is Hausdorff and definable by a countable family of seminorms, it is metrizable.

Chapter 2

General Topology

In mathematics, general topology or point-set topology is the branch of topology which studies properties of topological spaces and structures defined on them. It is distinct from other branches of topology in that the topological spaces may be very general, and do not have to be at all similar to manifolds. General topology provides the most general framework where fundamental concepts of topology such as open/closed sets, continuity, interior/exterior/boundary points, and limit points could be defined.

Definition

A topology is a pair (X,Σ) consisting of a set X and a collection Σ of subsets of X, called open sets, satisfying the following three axioms:

1. The union of open sets is an open set.
2. The finite intersection of open sets is an open set.
3. X and the empty set $\varnothing$ are open sets.

History

General topology grew out of a number of areas, most importantly the following:

- the detailed study of subsets of the real line (once known as the *topology of point sets*, this usage is now obsolete)
- the introduction of the manifold concept
- the study of metric spaces, especially normed linear spaces, in the early days of functional analysis.

General topology assumed its present form around 1940. It captures, one might say, almost everything in the intuition of continuity, in a technically adequate form that can be applied in any area of mathematics.

Scope

More specifically, it is in general topology that basic notions are defined and theorems about them proved. This includes the following:

- open and closed sets;
- interior and closure;
- neighbourhood and closeness;
- compactness and connectedness;
- continuous functions;
- convergence of sequences, nets, and filters;
- separation axioms
- countability axiom.

Other more advanced notions also appear, but are usually related directly to these fundamental concepts, without reference to other branches of mathematics. Set-theoretic topology examines such questions when they have substantial relations to set theory, as is often the case. Other main branches of topology are algebraic topology, geometric topology, and differential topology. As the name implies, general topology provides the common foundation for these areas.

An important variant of general topology is pointless topology, which, rather than using sets of points as its foundation, builds up topological concepts through the study of lattices, and, in particular, the category-theoretic study of frames and locales.

Open Set

Example: The points (x, y) satisfying $x^2 + y^2 = r^2$ are coloured blue. The points (x, y) satisfying $x^2 + y^2 < r^2$ are coloured red. The red points form an open set. The blue points form a closed set. The union of the red and blue points is a closed set.

In topology, a set U is called an open set if it is a neighbourhood of every point $x \in U$.[1] When dealing with metric spaces, this can be intuitively interpreted as saying that every $x \in U$ can be "moved" some non-zero distance, in any direction, and it will still lie within U.

The notion of an open set provides a fundamental way to speak of nearness of points in a topological space, without explicitly having a concept of distance defined. Concepts that use notions of nearness, such as the continuity of functions, can be translated into the language of open sets.

In point-set topology, open sets are used to distinguish between points and subsets of a space. The degree to which any two points can be separated is specified by the separation axioms. The collection of all open sets in a space defines the topology of the space.

Functions from one topological space to another that preserve the topology are the continuous functions. Although open sets and the topologies that they comprise are of central importance in point-set topology, they are also used as an organizational tool in other important branches of mathematics.

Examples of topologies include the Zariski topology in algebraic geometry that reflects the algebraic nature of varieties, and the topology on a differential manifold in differential topology where each point within the space is contained in an open set that is homeomorphic to an open ball in a finite-dimensional Euclidean space.

Point-set topology is the area of mathematics concerned with general topological spaces, and the relations between them. In the category of topological spaces, morphisms are continuous functions between topological spaces.

Continuous functions are readily observed to preserve topological structure, as they map "points close together" to "points close together"; that is, they preserve the structure of open sets defined on the space.

In metric topology, one can concretely define a distance function between two points, and thus metric spaces also have a topology, i.e. a certain structure of open sets defined on them. Thus, as opposed to the pure topological invariants, metric topology deals with isometries and the like; that is, distance preserving maps. In this case, the idea of an open set is used as an organizational tool rather than an object of study. From the topological point of view, metric spaces are fairly well understood, although many open problems still remain in metrizability theory.

Motivation

Intuitively, an open set provides a method to distinguish two points. For example, if about one point in a topological space there exists an open set not containing another (distinct) point, the two points are referred to as topologically distinguishable. In this manner, one may speak of whether two subsets of a topological space are "near" without concretely defining a metric on the topological space. Therefore, topological spaces may be seen as a generalization of metric spaces.

In the set of all real numbers, one has the natural Euclidean metric; that is, a function which measures the distance between two real numbers: $d(x, y) = |x - y|$. Therefore, given a real number, one can speak of the set of all points close to that real number; that is, within ε of that real number (refer to this real number as x).

In essence, points within ε of x approximate x to an accuracy of degree ε. Note that $\varepsilon > 0$ always but as ε becomes smaller and smaller, one obtains points that approximate x to a higher and higher degree of accuracy.

For example, if $x = 0$ and $\varepsilon = 1$, the points within ε of x are precisely the points of the interval (-1, 1); that is, the set of all real numbers between -1 and 1. However, with $\varepsilon = 0.5$, the points within ε of x are precisely the points of (-0.5, 0.5). Clearly, these points approximate x to a greater degree of accuracy compared to when $\varepsilon = 1$.

The previous discussion shows, for the case $x = 0$, that one may approximate x to higher and higher degrees of accuracy by defining ε to be smaller and smaller. In particular, sets of the form $(-\varepsilon, \varepsilon)$ give us a lot of information about points close to $x = 0$. Thus, rather than speaking of a concrete Euclidean metric, one may use sets to describe points close to x.

This innovative idea has far-reaching consequences; in particular, by defining different collections of sets containing 0 (distinct from the sets $(-\varepsilon, \varepsilon)$), one may find different results regarding the distance between 0 and other real numbers.

For example, if we were to define R as the only such set for "measuring distance", all points are close to 0 since there is only one possible degree of accuracy one may achieve in approximating 0: being a member of R. Thus, we find that in some sense, every real number is distance 0 away from ε It may help in this case to think of the measure as being a binary condition, all things in R are equally close to 0, while any item that is not in R is not close to 0.

In general, one refers to the family of sets containing 0, used to approximate 0, as a neighbourhood basis; a member of this neighbourhood basis is referred to as an open set. In fact, one may generalize these notions to an arbitrary set (X); rather than just the real numbers. In this case, given a point (x) of that set, one may define a collection of sets "around" (that is, containing) x, used to approximate x. Of course, this collection would have to satisfy certain properties (known as axioms) for otherwise we may not have a well-defined method to measure distance. For example, every point in X should approximate x to *some* degree of accuracy. Thus X should be in this family. Once we begin to define "smaller" sets containing x, we tend to approximate x to a greater degree of accuracy. Bearing this in mind, one may define the remaining axioms that the family of sets about x is required to satisfy.

Definitions

The concept of open sets can be formalized with various degrees of generality, for example:

Euclidean Space

A subset U of the Euclidean n-space R^n is called *open* if, given any point x in U, there exists a real number $\varepsilon > 0$ such that, given any point y in R^n whose Euclidean distance from x is smaller than ε, y also belongs to U. Equivalently, a subset U of R^n is open if every point in U has a neighbourhood in R^n contained in U.

Metric Spaces

A subset U of a metric space (M, d) is called *open* if, given any point x in U, there exists a real number $\varepsilon > 0$ such that, given any point y in M with $d(x, y) < \varepsilon$, y also belongs to U. Equivalently, U is open if every point in U has a neighbourhood contained in U.

This generalises the Euclidean space example, since Euclidean space with the Euclidean distance is a metric space.

Topological Spaces

If (X, 5Ø) is a topological space, then a subset O of X is called *open* if and only if O is a neighbourhood of each of its points.

Note that infinite intersections of open sets need not be open. For example, the intersection of all intervals of the form $(-1/n, 1/n)$, where n is a positive integer, is the set {0} which is not open in the real line. Sets that can be constructed as the intersection of countably many open sets are denoted G_δ sets.

The topological definition of open sets generalises the metric space definition: If one begins with a metric space and defines open sets as before, then the family of all open sets is a topology on the metric space. Every metric space is therefore, in a natural way, a topological space. There are, however, topological spaces that are not metric spaces.

Properties

- The empty set is both open and closed (clopen set).
- The set X that the topology is defined on is both open and closed (clopen set).
- The union of any number of open sets is open.
- The intersection of a finite number of open sets is open.

Uses

Open sets have a fundamental importance in topology. The concept is required to define and make sense of topological space and other topological structures that deal with the notions of closeness and convergence for spaces such as metric spaces and uniform spaces.

Every subset A of a topological space X contains a (possibly empty) open set; the largest such open set is called the interior of A. It can be constructed by taking the union of all the open sets contained in A. Given topological spaces X and Y, a function f from X to Y is *continuous* if the preimage of every open set in Y is open in X. The function f is called *open* if the image of every open set in X is open in Y. An open set on the real line has the characteristic property that it is a countable union of disjoint open intervals.

Notes and Cautions

"Open" is Defined Relative to a Particular Topology: Whether a set is open depends on the topology under consideration. Having opted for greater brevity over greater clarity, we refer to a set X endowed with a topology T as "the topological space X" rather than "the topological space (X, T)", despite the fact that all the topological data is contained in T.

If there are two topologies on the same set, a set U that is open in the first topology might fail to be open in the second topology. For example, if X is any topological space and Y is any subset of X, the set Y can be given its own topology (called the 'subspace topology') defined by "a set U is open in the subspace topology on Y if and only if U is the intersection of Y with an open set from the original topology on X." This potentially introduces new open sets: if V is open in the original topology on X, but isn't, then is open in the subspace topology on Y but $V \cap Y$ not $V \cap Y$ in the original topology on X.

As a concrete example of this, if U is defined as the set of rational numbers in the interval (0, 1), then U is an open subset of the rational numbers, but not of the real numbers. This is because when the surrounding space is the rational numbers, for every point x in U, there exists a positive number a such that all *rational* points within distance a of x are also in U. On the other hand, when the surrounding space is the reals, then for every point x in U there is *no* positive a such that all *real* points within distance a of x are in U (since U contains no non-rational numbers).

Closed Set

In geometry, topology, and related branches of mathematics, a closed set is a set whose complement is an open set. In a topological space, a closed set can be defined as a set which contains all its limit points. In a complete metric space, a closed set is a set which is closed under the limit operation.

Equivalent Definitions of a Closed Set

In a topological space, a set is closed if and only if it coincides with its closure. Equivalently, a set is closed if and only if it contains all of its limit points.

This is not to be confused with a closed manifold.

Properties of Closed Sets

A closed set contains its own boundary. In other words, if you are "outside" a closed set, you may move a small amount in any direction and still stay outside the set. Note that this is also true if the boundary is the empty set, e.g. in the metric space of rational numbers, for the set of numbers of which the square is less than 2.

Any intersection of closed sets is closed (including intersections of infinitely many closed sets), and any union of *finitely many* closed sets is closed. In particular, the empty set and the whole space are closed. In fact, given a set X and a collection F of subsets of X that has these properties, then F will be the collection of closed sets for a unique topology on X.

The intersection property also allows one to define the closure of a set A in a space X, which is defined as the smallest closed subset of X that is a superset of A. Specifically, the closure of A can be constructed as the intersection of all of these closed supersets.

Sets that can be constructed as the union of countably many closed sets are denoted F_σ sets. These sets need not be closed.

Examples of Closed Sets

- The closed interval $[a,b]$ of real numbers is closed for an explanation of the bracket and parenthesis set notation.)
- The unit interval [0,1] is closed in the metric space of real numbers, and the set $[0,1] \cap Q$ of rational numbers between 0 and 1 (inclusive) is closed in the space of rational numbers, but $[0,1] \cap Q$ is not closed in the real numbers.
- Some sets are neither open nor closed, for instance the half-open interval [0,1) in the real numbers.
- Some sets are both open and closed and are called clopen sets.
- Half-interval $[1,\infty)$ is closed.
- The Cantor set is an unusual closed set in the sense that it consists entirely of boundary points and is nowhere dense.
- Singleton points (and thus finite sets) are closed in Hausdorff spaces.
- If X and Y are topological spaces, a function f from X into Y is continuous if and only if preimages of closed sets in Y are closed in X.

More About Closed Sets

In point set topology, a set A is closed if it contains all its boundary points.

The notion of closed set is defined above in terms of open sets, a concept that makes sense for topological spaces, as well as for other spaces that carry topological structures, such as metric spaces, differentiable manifolds, uniform spaces, and gauge spaces.

An alternative characterization of closed sets is available via sequences and nets. A subset A of a topological space X is closed in X if and only if every limit of every net of elements of A also belongs to A. In a first-countable space (such as a metric space), it is enough to consider only convergent sequences, instead of all nets. One value of this characterisation is that it may be used as a definition in the context of convergence spaces, which are more general than topological spaces. Notice that this characterisation also depends on the surrounding space *X*, because whether or not a sequence or net converges in *X* depends on what points are present in *X*.

Whether a set is closed depends on the space in which it is embedded. However, the compact Hausdorff spaces are "absolutely closed", in the sense that, if you embed a compact Hausdorff space *K* in an arbitrary Hausdorff space *X*, then *K* will always be a closed subset of *X*; the "surrounding space" does not matter here. Stone-Èech compactification, a process that turns a completely regular Hausdorff space into a compact Hausdorff space, may be described as adjoining limits of certain nonconvergent nets to the space.

Furthermore, every closed subset of a compact space is compact, and every compact subspace of a Hausdorff space is closed. Closed sets also give a useful characterization of compactness: a topological space X is compact if and only if every collection of nonempty closed subsets of X with empty intersection admits a finite subcollection with empty intersection.

A topological space X is disconnected if there exist disjoint, nonempty, closed subsets A and B of X whose union is X. Furthermore, X is totally disconnected if it has an open basis consisting of closed sets.

Interior (Topology)

In mathematics, specifically in topology, the interior of a set *S* of points of a topological space consists of all points of *S* that do not belong to the boundary of *S*. A point that is in the interior of *S* is an interior point of *S*.

Equivalently the interior of *S* is the complement of the closure of the complement of *S*. In this sense interior and closure are dual notions.

The exterior of a set is the interior of its complement, equivalently the complement of its closure; it consists of the points that are in neither the set nor its boundary. The interior, boundary, and exterior of a subset together partition the whole space into three blocks (or fewer when one or more of these is empty).

The interior and exterior are always open while the boundary is always closed. Sets with empty interior are often called hollow.

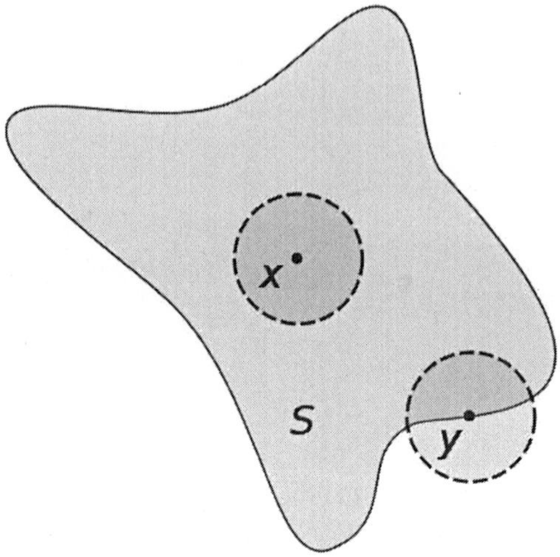

Figure: *The point* x *is an interior point of* S, *since it is contained within* S *together with an open ball around it. The point* y *is on the boundary of* S.

Definitions

Interior Point: If S is a subset of a Euclidean space, then x is an interior point of S if there exists an open set centered at x which is contained in S.

This definition generalizes to any subset S of a metric space X. Fully expressed, if X is a metric space with metric d, then x is an interior point of S if there exists $r > 0$, such that y is in S whenever the distance $d(x, y) < r$.

This definition generalises to topological spaces by replacing "open ball" with "neighbourhood". Let S be a subset of a topological space X. Then x is an interior point of S if there exists a neighbourhood of x which is contained in S. Note that this definition does not depend upon whether neighbourhoods are required to be open.

If neighbourhoods are not required to be open then S will automatically be a neighbourhood of x if S contains a neighbourhood of x.

Interior of a Set

The interior of a set S is the set of all interior points of S. The interior of S is denoted int(S), Int(S), or S°. The interior of a set has the following properties.

- int(S) is an open subset of S.
- int(S) is the union of all open sets contained in S.
- int(S) is the largest open set contained in S.
- A set S is open if and only if S = int(S).
- int(int(S)) = int(S) (idempotence).
- If S is a subset of T, then int(S) is a subset of int(T).
- If A is an open set, then A is a subset of S if and only if A is a subset of int(S).

Sometimes the second or third property above is taken as the *definition* of the topological interior.

Note that these properties are also satisfied if "interior", "subset", "union", "contained in", "largest" and "open" are replaced by "closure", "superset", "intersection", "which contains", "smallest", and "closed", respectively. For more on this matter.

Interior Operator

The interior operator $^{\circ}$ is dual to the closure operator $^{-}$, in the sense that

$$S^{\circ} = X \setminus (X \setminus S)^{-},$$

and also

$$S^{-} = X \setminus (X \setminus S)^{\circ}$$

where X is the topological space containing S, and the backslash refers to the set-theoretic difference.

Therefore, the abstract theory of closure operators and the Kuratowski closure axioms can be easily translated into the language of interior operators, by replacing sets with their complements.

Exterior of a Set

The exterior of a subset S of a topological space X, denoted ext(S) or Ext(S), is the interior int($X \setminus S$) of its relative complement. Alternatively, it can be defined as $X \setminus S^{-}$, the complement of the closure of S. Many properties follow in a straightforward way from those of the interior operator, such as the following.

- ext(S) is an open set that is disjoint with S.
- ext(S) is the union of all open sets that are disjoint with S.
- ext(S) is the largest open set that is disjoint with S.
- If S is a subset of T, then ext(S) is a superset of ext(T).

Unlike the interior operator, ext is not idempotent, but the following holds:

- ext(ext(S)) is a superset of int(S).

Closure (Topology)

In mathematics, the closure of a subset S in a topological space consists of all points in S plus the limit points of S. Intuitively, these are all the points that are "near" S. A point which is in the closure of S is a point of closure of S. The notion of closure is in many ways dual to the notion of interior.

Definitions

Point of Closure: For S a subset of a Euclidean space, x is a point of closure of S if every open ball centered at x contains a point of S (this point may be x itself).

This definition generalises to any subset S of a metric space X. Fully expressed, for X a metric space with metric d, x is a point of closure of S if for every $r > 0$, there is a y in S such that the distance $d(x, y) < r$. (Again, we may have $x = y$.) Another way to express this is to say that x is a point of closure of S if the distance $d(x, S) := \inf\{d(x, s) : s \text{ in } S\} = 0$.

This definition generalises to topological spaces by replacing "open ball" or "ball" with "neighbourhood". Let S be a subset of a topological space X. Then x is a point of closure of S if every neighbourhood of x contains a point of S. Note that this definition does not depend upon whether neighbourhoods are required to be open.

Limit Point

The definition of a point of closure is closely related to the definition of a limit point. The difference between the two definitions is subtle but important — namely, in the definition of limit point, every neighbourhood of the point x in question must contain a point of the set *other than x itself.*

Thus, every limit point is a point of closure, but not every point of closure is a limit point. A point of closure which is not a limit point is an isolated point.

In other words, a point x is an isolated point of S if it is an element of S and if there is a neighbourhood of x which contains no other points of S other than x itself.

For a given set S and point x, x is a point of closure of S if and only if x is an element of S or x is a limit point of S (or both).

Closure of a Set

The closure of a set S is the set of all points of closure of S. The closure of S is denoted cl(S), Cl(S), $\overline{S}$ or $\overline{S}$. The closure of a set has the following properties.

- cl(S) is a closed superset of S.
- cl(S) is the intersection of all closed sets containing S.
- cl(S) is the smallest closed set containing S.
- A set S is closed if and only if S = cl(S).
- If S is a subset of T, then cl(S) is a subset of cl(T).
- If A is a closed set, then A contains S if and only if A contains cl(S).

Sometimes the second or third property above is taken as the *definition* of the topological closure, which still make sense when applied to other types of closures.

In a first-countable space (such as a metric space), cl(S) is the set of all limits of all convergent sequences of points in S. For a general topological space, this statement remains true if one replaces "sequence" by "net" or "filter".

Note that these properties are also satisfied if "closure", "superset", "intersection", "contains/containing", "smallest" and "closed" are replaced by "interior", "subset", "union", "contained in", "largest", and "open". For more on this matter.

Connected Space

Connected and disconnected subspaces of R^2

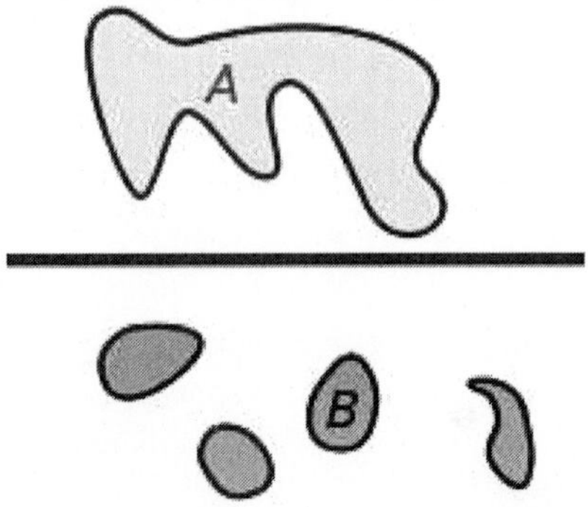

Figure: *The green space* A *at top is simply connected whereas the blue space* B *below is not connected.*

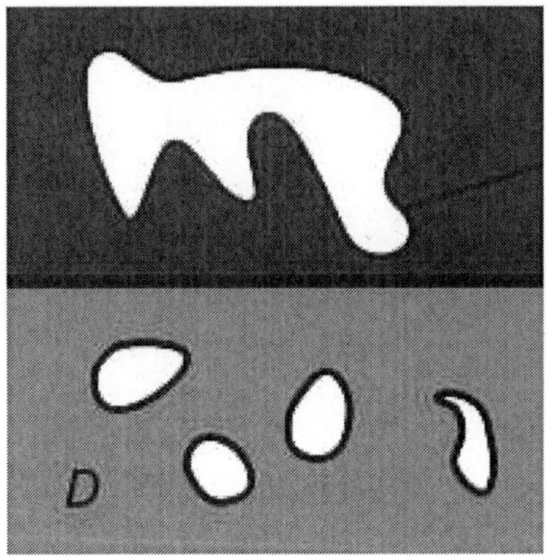

Figure: *The pink space* C *at top and the orange space* D *are both connected;* C *is also simply connected while* D *is not.*

In topology and related branches of mathematics, a connected space is a topological space that cannot be represented as the union of two or more disjoint nonempty open subsets. Connectedness is one of the principal topological properties that is used to distinguish topological spaces. A stronger notion is that of a path-connected space, which is a space where any two points can be joined by a path.

A subset of a topological space X is a connected set if it is a connected space when viewed as a subspace of X. As an example of a space that is not connected, one can delete an infinite line from the plane. Other examples of disconnected spaces (that is, spaces which are not connected) include the plane with a closed annulus removed, as well as the union of two disjoint open disks in two-dimensional Euclidean space.

Formal Definition

A topological space X is said to be disconnected if it is the union of two disjoint nonempty open sets. Otherwise, X is said to be connected. A subset of a topological space is said to be connected if it is connected under its subspace topology. Some authors exclude the empty set (with its unique topology) as a connected space, but this article does not follow that practice.

For a topological space X the following conditions are equivalent:

1. X is connected.
2. X cannot be divided into two disjoint nonempty closed sets.
3. The only subsets of X which are both open and closed (clopen sets) are X and the empty set.
4. The only subsets of X with empty boundary are X and the empty set.
5. X cannot be written as the union of two nonempty separated sets.
6. The only continuous functions from X to $\{0,1\}$ are constant.

Connected Components

The maximal connected subsets (ordered by inclusion) of a nonempty topological space are called the connected components of the space. The components of any topological space X form a partition of X: they are disjoint, nonempty, and their union is the whole space.

Every component is a closed subset of the original space. It follows that, in the case where their number is finite, each component is also an open subset. However, if their number is infinite, this might not be the case; for instance, the connected components of the set of the rational numbers are the one-point sets, which are not open.

Let Γ_x be the connected component of x in a topological space X, and $\Gamma_{x'}$ be the intersection of all open-closed sets containing x (called quasi-component of x.) Then $\Gamma_x \subset \Gamma'_x$ where the equality holds if X is compact Hausdorff or locally connected.

Disconnected Spaces

A space in which all components are one-point sets is called totally disconnected. Related to this property, a space X is called totally separated if, for any two distinct elements x and y of X, there exist disjoint open neighbourhoods U of x and V of y such that X is the union of U and V.

Clearly any totally separated space is totally disconnected, but the converse does not hold. For example take two copies of the rational numbers Q, and identify them at every point except zero.

The resulting space, with the quotient topology, is totally disconnected. However, by considering the two copies of zero, one sees that the space is not totally separated. In fact, it is not even Hausdorff, and the condition of being totally separated is strictly stronger than the condition of being Hausdorff.

Examples

- The closed interval [0, 2] in the standard subspace topology is connected; although it can, for example, be written as the union of [0, 1) and [1, 2], the second set is not open in the aforementioned topology of [0, 2].
- The union of [0, 1) and (1, 2] is disconnected; both of these intervals are open in the standard topological space $[0, 1) \cup (1, 2]$.
- $(0, 1) \cup \{3\}$ is disconnected.
- A convex set is connected; it is actually simply connected.
- A Euclidean plane excluding the origin, (0, 0), is connected, but is not simply connected. The three-dimensional Euclidean space without the origin is connected, and even simply connected. In contrast, the one-dimensional Euclidean space without the origin is not connected.

- A Euclidean plane with a straight line removed is not connected since it consists of two half-planes.
- The space of real numbers with the usual topology is connected.
- Any topological vector space over a connected field is connected.
- Every discrete topological space with at least two elements is disconnected, in fact such a space is totally disconnected. The simplest example is the discrete two-point space.
- On the other hand, a finite set might be connected. For example, the spectrum of a discrete valuation ring consists of two points and is connected. It is an example of a Sierpiñski space.
- The Cantor set is totally disconnected; since the set contains uncountably many points, it has uncountably many components.
- If a space X is homotopy equivalent to a connected space, then X is itself connected.
- The topologist's sine curve is an example of a set that is connected but is neither path connected nor locally connected.
- The general linear group $\mathrm{GL}(n, \mathbf{R})$ (that is, the group of n-by-n real matrices) consists of two connected components: the one with matrices of positive determinant and the other of negative determinant. In particular, it is not connected. In contrast, $\mathrm{GL}(n, \mathbf{C})$ is connected. More generally, the set of invertible bounded operators on a (complex) Hilbert space is connected.
- The spectra of commutative local ring and integral domains are connected. More generally, the following are equivalent

1. The spectrum of a commutative ring R is connected
2. Every finitely generated projective module over R has constant rank.
3. R has no idempotent $\neq 0,1$ (i.e., R is not a product of two rings in a nontrivial way).

Path Connectedness

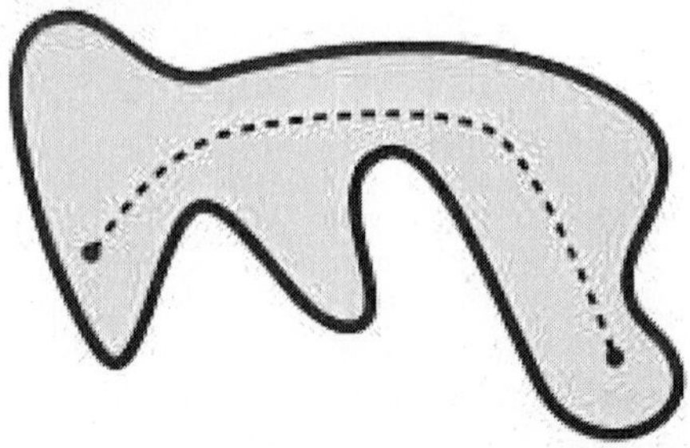

Figure: *This subspace of R^2 is path-connected, because a path can be drawn between any two points in the space.*

A path from a point x to a point y in a topological space X is a continuous function f from the unit interval [0,1] to X with $f(0) = x$ and $f(1) = y$. A path-component of X is an equivalence class of X under the equivalence relation which makes x equivalent to y if there is a path from x to y.

The space X is said to be path-connected (or pathwise connected or 0-connected) if there is at most one path-component, i.e. if there is a path joining any two points in X. Again, many others exclude the empty space. Every path-connected space is connected. The converse is not always true: examples of connected spaces that are not path-connected include the extended long line L^* and the *topologist's sine curve.*

However, subsets of the real line R are connected if and only if they are path-connected; these subsets are the intervals of R. Also, open subsets of R^n or C^n are connected if and only if they are path-connected. Additionally, connectedness and path-connectedness are the same for finite topological spaces.

Arc Connectedness

A space X is said to be arc-connected or arcwise connected if any two distinct points can be joined by an *arc*, that is a path f which is a homeomorphism between the unit interval [0, 1] and its image $f([0, 1])$. It can be shown any Hausdorff space which is path-connected is also arc-connected.

An example of a space which is path-connected but not arc-connected is provided by adding a second copy 0' of 0 to the nonnegative real numbers $[0,\infty)$. One endows this set with a partial order by specifying that $0' < a$ for any positive number a, but leaving 0 and 0' incomparable.

One then endows this set with the *order topology*, that is one takes the open intervals $(a, b) = \{x \mid a < x < b\}$ and the half-open intervals $[0, a) = \{x \mid 0 \leq x < a\}$, $[0', a) = \{x \mid 0' \leq x < a\}$ as a base for the topology. The resulting space is a T_1 space but not a Hausdorff space. Clearly 0 and 0' can be connected by a path but not by an arc in this space.

Local Connectedness

A topological space is said to be locally connected at a point x if every neighbourhood of x contains a connected open neighbourhood. It is locally connected if it has a base of connected sets. It can be shown that a space X is locally connected if and only if every component of every open set of X is open. The topologist's sine curve is an example of a connected space that is not locally connected.

Similarly, a topological space is said to be locally path-connected if it has a base of path-connected sets. An open subset of a locally path-connected space is connected if and only if it is path-connected. This generalizes the earlier statement about R^n and C^n, each of which is locally path-connected. More generally, any topological manifold is locally path-connected.

Theorems

- *Main theorem*: Let X and Y be topological spaces and let $f: X \to Y$ be a continuous function. If X is (path-)connected then the image $f(X)$ is (path-)connected. This result can be considered a generalization of the intermediate value theorem.
- If $\{A_i\}_{i \in I}$ is a family of connected subsets of a topological space X indexed by an arbitrary set I such that for all i, J in I, $A_i \cap A_j$ is nonempty, then $\cup A_i$ is also connected.
- If $\{A_\alpha\}$ is a nonempty family of connected subsets of a topological space X such that $\cap A_\alpha$ is nonempty, then $\cap A_\alpha$ is also connected.
- Every path-connected space is connected.

- Every locally path-connected space is locally connected.
- A locally path-connected space is path-connected if and only if it is connected.
- The closure of a connected subset is connected.
- The connected components are always closed (but in general not open)
- The connected components of a locally connected space are also open.
- The connected components of a space are disjoint unions of the path-connected components (which in general are neither open nor closed).
- Every quotient of a connected (resp. path-connected) space is connected (resp. path-connected).
- Every product of a family of connected (resp. path-connected) spaces is connected (resp. path-connected).
- Every open subset of a locally connected (resp. locally path-connected) space is locally connected (resp. locally path-connected).
- Every manifold is locally path-connected.

Graphs

Graphs have path connected subsets, namely those subsets for which every pair of points has a path of edges joining them. But it is not always possible to find a topology on the set of points which induces the same connected sets. The 5-cycle graph (and any n-cycle with $n>3$ odd) is one such example.

As a consequence, a notion of connectedness can be formulated independently of the topology on a space. To wit, there is a category of connective spaces consisting of sets with collections of connected subsets satisfying connectivity axioms; their morphisms are those functions which map connected sets to connected sets (Muscat & Buhagiar 2006).

Topological spaces and graphs are special cases of connective spaces; indeed, the finite connective spaces are precisely the finite graphs. However, every graph can be canonically made into a topological space, by treating vertices as points and edges as copies of the unit interval. Then one can show that the graph is connected (in the graph theoretical sense) if and only if it is connected as a topological space.

Stronger Forms of Connectedness

There are stronger forms of connectedness for topological spaces, for instance:

- If there exist no two disjoint non-empty open sets in a topological space, X, X must be connected, and thus hyperconnected spaces are also connected.
- Since a simply connected space is, by definition, also required to be path connected, any simply connected space is also connected. Note however, that if the "path connectedness" requirement is dropped from the definition of simple connectivity, a simply connected space does not need to be connected.
- Yet stronger versions of connectivity include the notion of a contractible space. Every contractible space is path connected and thus also connected.

In general, note that any path connected space must be connected but there exist connected spaces that are not path connected. The deleted comb space furnishes such an example, as does the above mentioned topologist's sine curve.

Defining Topologies Via Continuous Functions

Given a function:

$$f : X \to S,$$

where X is a topological space and S is a set (without a specified topology), the final topology on S is defined by letting the open sets of S be those subsets A of S for which $f^{1}(A)$ is open in X. If S has an existing topology, f is continuous with respect to this topology if and only if the existing topology is coarser than the final topology on S. Thus the final topology can be characterized as the finest topology on S that makes f continuous. If f is surjective, this topology is canonically identified with the quotient topology under the equivalence relation defined by f.

Dually, for a function f from a set S to a topological space, the initial topology on S has as open subsets A of S those subsets for which $f(A)$ is open in X. If S has an existing topology, f is continuous with respect to this topology if and only if the existing topology is finer than the initial topology on S. Thus the initial topology can be characterized as the coarsest topology on S that makes f continuous. If f is injective, this topology is canonically identified with the subspace topology of S, viewed as a subset of X. More generally, given a set S, specifying the set of continuous functions

$$S \to X$$

into all topological spaces X defines a topology. Dually, a similar idea can be applied to maps

$$X \to S.$$

This is an instance of a universal property.

Related notions

Various other mathematical domains use the concept of continuity in different, but related meanings. For example, in order theory, an order-preserving function $f: X \to Y$ between two complete lattices X and Y (particular types of partially ordered sets) is continuous if for each subset A of X, we have $\sup(f(A)) = f(\sup(A))$. Here sup is the supremum with respect to the orderings in X and Y, respectively. Applying this to the complete lattice consisting of the open subsets of a topological space, this gives back the notion of continuity for maps between topological spaces.

In category theory, a functor

$$F : \mathcal{C} \to \mathcal{D}$$

between two categories is called *continuous*, if it commutes with small limits. That is to say,

$$\varprojlim_{i \in I} F(C_i) \cong F(\varprojlim_{i \in I} C_i)$$

for any small (i.e., indexed by a set I, as opposed to a class) diagram of objects in C.

A *continuity space* is a generalization of metric spaces and posets, which uses the concept of quantales, and that can be used to unify the notions of metric spaces and domains.

Net (Mathematics)

In mathematics, more specifically in general topology and related branches, a net or Moore–Smith sequence is a generalization of the notion of a sequence. In essence, a sequence is a function with domain the natural numbers, and in the context of topology, the range of this function is usually any topological space. However, in the context of topology, sequences do not fully encode all information about a function between topological spaces. In particular, the following two conditions are not equivalent in general for a map f between topological spaces X and Y:

1. The map f is continuous
2. Given any point x in X, and any sequence in X converging to x, the composition of f with this sequence converges to $f(x)$

It is true, however, that condition 1 implies condition 2 in the context of all spaces. The difficulty encountered when attempting to prove that condition 2 implies condition 1 lies in the fact that topological spaces are, in general, not first-countable. If the first-countability axiom were imposed on the topological spaces in question, the two above conditions would be equivalent. In particular, the two conditions are equivalent for metric spaces.

The purpose of the concept of a net, first introduced by E. H. Moore and H. L. Smith in 1922, is to generalize the notion of a sequence so as to confirm the equivalence of the conditions (with "sequence" being replaced by "net" in condition 2). In particular, rather than being defined on a countable linearly ordered set, a net is defined on an arbitrary directed set.

In particular, this allows theorems similar to that asserting the equivalence of condition 1 and condition 2, to hold in the context of topological spaces which do not necessarily have a countable or linearly ordered neighbourhood basis around a point.

Therefore, while sequences do not encode sufficient information about functions between topological spaces, nets do because collections of open sets in topological spaces are much like directed sets in behaviour. The term "net" was coined by Kelley. Nets are one of the many tools used in topology to generalize certain concepts that may only be general enough in the context of metric spaces. A related notion, that of the filter, was developed in 1937 by Henri Cartan.

Definition

If X is a topological space, a *net* in X is a function from some directed set A to X.

If A is a directed set, we often write a net from A to X in the form (x_α), which expresses the fact that the element α in A is mapped to the element x_α in X.

Examples of Nets

Every non-empty totally ordered set is directed. Therefore every function on such a set is a net. In particular, the natural numbers with the usual order form such a set, and a sequence is a function on the natural numbers, so every sequence is a net.

Another important example is as follows. Given a point x in a topological space, let N_x denote the set of all neighbourhoods containing x. Then N_x is a directed set, where the direction is given by reverse inclusion, so that $S \geq T$ if and only if S is contained in T. For S in N_x, let x_S be a point in S. Then (x_S) is a net. As S increases with respect to $\geq$, the points x_S in the net are constrained to lie in decreasing neighbourhoods of x, so intuitively speaking, we are led to the idea that x_S must tend towards x in some sense. We can make this limiting concept precise.

Limits of Nets

If (x_α) is a net from a directed set A into X, and if Y is a subset of X, then we say that (x_α) is eventually in Y (or residually in Y) if there exists an α in A so that for every β in A with $\beta \geq \alpha$, the point x_β lies in Y.

If (x_α) is a net in the topological space X, and x is an element of X, we say that the net converges towards x or has limit x and write

$\lim x_\alpha = x$

if and only if

for every neighbourhood U of x, (x_α) is eventually in U.

Intuitively, this means that the values x_α come and stay as close as we want to x for large enough α.

Note that the example net given above on the neighbourhood system of a point x does indeed converge to x according to this definition.

Given a base for the topology, in order to prove convergence of a net it is necessary and sufficient to prove that there exists some point x, such that (x_α) is eventually in all members of the base containing this putative limit.

Examples of Limits of Nets

- Limit of a sequence and limit of a function.
- Limits of nets of Riemann sums, in the definition of the Riemann integral. In this example, the directed set is the set of partitions of the interval of integration, partially ordered by inclusion.

Supplementary Definitions

Let φ be a net on X based on the directed set D and let A be a subset of X, then φ is said to be frequently in (or cofinally in) A if for every α in D there exists some $\beta \geq \alpha$, β in D, so that $\varphi(\beta)$ is in A.

A point x in X is said to be an accumulation point or cluster point of a net if (and only if) for every neighbourhood U of x, the net is frequently in U.

A net φ on set X is called universal, or an ultranet if for every subset A of X, either φ is eventually in A or φ is eventually in X-A.

Examples

Sequence in a topological space:

A sequence $(a_1, a_2, ...)$ in a topological space V can be considered a net in V defined on N.

The net is eventually in a subset Y of V if there exists an N in N such that for every $n \geq N$, the point a_n is in Y.

We have $\lim_{x \to c} a_n = L$ if and only if for every neighbourhood Y of L, the net is eventually in Y.

The net is frequently in a subset Y of V if and only if for every N in N there exists some $n \geq N$ such that a_n is in Y, that is, if and only if infinitely many elements of the sequence are in Y. Thus a point y in V is a cluster point of the net if and only if every neighbourhood Y of y contains infinitely many elements of the sequence.

Function from a metric space to a topological space:

Consider a function from a metric space M to a topological space V, and a point c of M. We direct the set $M\setminus\{c\}$ reversely according to distance from c, that is, the relation is "has at least the same distance to c as", so that "large enough" with respect to the relation means "close enough to c". The function f is a net in V defined on $M\setminus\{c\}$.

The net f is eventually in a subset Y of V if there exists an a in $M\setminus\{c\}$ such that for every x in $M\setminus\{c\}$ with $d(x,c) \leq d(a,c)$, the point $f(x)$ is in Y.

We have $\lim_{x \to c} f(x) = L$ if and only if for every neighbourhood Y of L, f is eventually in Y.

The net f is frequently in a subset Y of V if and only if for every a in $M\setminus\{c\}$ there exists some x in $M\setminus\{c\}$ with $d(x,c) \leq d(a,c)$ such that $f(x)$ is in Y. A point y in V is a cluster point of the net f if and only if for every neighbourhood Y of y, the net is frequently in Y.

Function from a well-ordered set to a topological space:

Consider a well-ordered set $[0, c]$ with limit point c, and a function f from $[0, c)$ to a topological space V. This function is a net on $[0, c)$.

It is eventually in a subset Y of V if there exists an a in $[0, c)$ such that for every $x \geq a$, the point $f(x)$ is in Y.

We have $\lim_{x \to c} f(x) = L$ if and only if for every neighbourhood Y of L, f is eventually in Y.

The net f is frequently in a subset Y of V if and only if for every a in $[0, c)$ there exists some x in $[a, c)$ such that $f(x)$ is in Y.

A point y in V is a cluster point of the net f if and only if for every neighbourhood Y of y, the net is frequently in Y.

The first example is a special case of this with $c = \Omega$.

Properties

Virtually all concepts of topology can be rephrased in the language of nets and limits. This may be useful to guide the intuition since the notion of limit of a net is very similar to that of limit of a sequence. The following set of theorems and lemmas help cement that similarity:

- A function $f:X \to Y$ between topological spaces is continuous at the point x if and only if for every net (x_α) with

$$\lim x_\alpha = x$$

we have

$$\lim f(x_\alpha) = f(x).$$

Note that this theorem is in general not true if we replace "net" by "sequence". We have to allow for more directed sets than just the natural numbers if X is not first-countable.

- In general, a net in a space X can have more than one limit, but if X is a Hausdorff space, the limit of a net, if it exists, is unique. Conversely, if X is not Hausdorff, then there exists a net on X with two distinct limits.

  Thus the uniqueness of the limit is *equivalent* to the Hausdorff condition on the space, and indeed this may be taken as the definition. Note that this result depends on the directed-

  ness condition; a set indexed by a general preorder or partial order may have distinct limit points even in a Hausdorff space.

- If U is a subset of X, then x is in the closure of U if and only if there exists a net (x_α) with limit x and such that x_α is in U for all α.
- A subset A of X is closed if and only if, whenever (x_α) is a net with elements in A and limit x, then x is in A.

The set of cluster points of a net is equal to the set of limits of its convergent subnets.

- A net has a limit if and only if all of its subnets have limits. In that case, every limit of the net is also a limit of every subnet.
- A space X is compact if and only if every net (x_α) in X has a subnet with a limit in X. This can be seen as a generalization of the Bolzano–Weierstrass theorem and Heine–Borel theorem.
- A net in the product space has a limit if and only if each projection has a limit. Symbolically, if (x_α) is a net in the product $X = \pi_i X_i$, *then it converges to* x *if and only*

if $\pi_i(x_\alpha) \to \pi_i(x)$ *for each* i. *Armed with this observation and the above characterization of compactness in terms on nets, one can give a slick proof of Tychonoff's theorem.*

- If $f{:}X \to Y$ and (x_α) is an ultranet on X, then $(f(x_\alpha))$ is an ultranet on Y.

Related Ideas

In a metric space or uniform space, one can speak of *Cauchy nets* in much the same way as Cauchy sequences. The concept even generalises to Cauchy spaces.

The theory of filters also provides a definition of *convergence* in general topological spaces.

Limit Superior

Limit superior and limit inferior of a net of real numbers can be defined in a similar manner as for sequences. Some authors work even with more general structures, like complete lattices.

For a net $(x_\alpha)_{\alpha \in I}$ we put

$$\limsup x_\alpha = \lim_{\alpha \in I} \sup_{\beta \succeq \alpha} x_\beta = \inf_{\alpha \in I} \sup_{\beta \succeq \alpha} x_\beta.$$

Limit superior of a net of real numbers has many properties analogous to the case of sequences, e.g.

$$\limsup(x_\alpha + y_\alpha) \leq \limsup x_\alpha + \limsup y_\alpha,$$

where equality holds whenever one of the nets is convergent.

Filter (Mathematics)

In mathematics, a filter is a special subset of a partially ordered set. A frequently used special case is the situation that the ordered set under consideration is just the power set of some set, ordered by set inclusion. Filters appear in order and lattice theory, but can also be found in topology whence they originate. The dual notion of a filter is an ideal. Filters were introduced by Henri Cartan in 1937 and subsequently used by Bourbaki in their book *Topologie Générale* as an alternative to the similar notion of a net developed in 1922 by E. H. Moore and H. L. Smith.

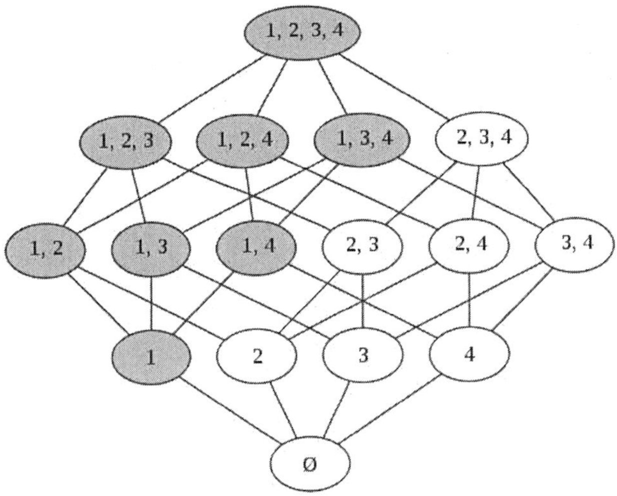

Figure: *The powerset algebra of the set* $\{1,2,3,4\}$ *with the upset* $\uparrow\{1\}$ *coloured green. The green elements make a* principal ultrafilter *on the lattice.*

General Definition

A non-empty subset *F* of a partially ordered set $(P,\leq)$ is a filter if the following conditions hold:

1. For every *x*, *y* in *F*, there is some element *z* in *F* such that $z \leq x$ and $z \leq y$. (*F* is a filter base)
2. For every *x* in *F* and *y* in *P*, $x \leq y$ implies that *y* is in *F*. (*F* is an *upper set*)
3. A filter is proper if it is not equal to the whole set *P*. This is sometimes omitted from the definition of a filter.

While the above definition is the most general way to define a filter for arbitrary posets, it was originally defined for lattices only. In this case, the above definition can be characterized by the following equivalent statement: A non-empty subset *F* of a lattice $(P,\leq)$ is a filter, if and only if it is an upper set that is closed under finite meets (infima), i.e., for all *x*, *y* in *F*, we find that $x \neq y$ is also in *F*.

The smallest filter that contains a given element *p* is a principal filter and *p* is a principal element in this situation.

The principal filter for p is just given by the set $\{x \text{ in } P \mid p \leq x\}$ and is denoted by prefixing p with an upward arrow: $\uparrow p$.

The dual notion of a filter, i.e. the concept obtained by reversing all $\leq$ and exchanging $\wedge$ with $\vee$, is ideal. Because of this duality, the discussion of filters usually boils down to the discussion of ideals. Hence, most additional information on this topic (including the definition of maximal filters and prime filters) is to be found in the article on ideals. There is a separate article on ultrafilters.

Filter on a Set

A special case of a filter is a filter defined on a set. Given a set S, a partial ordering $\subseteq$ can be defined on the powerset P(S) by subset inclusion, turning (P(S),$\subseteq$) into a lattice. Define a filter F on S as a subset of P(S) with the following properties:

1. S is in F. (*F is non-empty*)
2. The empty set is not in F. (*F is proper*)
3. If A and B are in F, then so is their intersection. (*F is closed under finite meets*)
4. If A is in F and A is a subset of B, then B is in F, for all subsets B of S. (*F is an upper set*)

The first three properties imply that a filter on a set has the finite intersection property. Note that with this definition, a filter on a set is indeed a filter; in fact, it is a proper filter. Because of this, sometimes this is called a proper filter on a set; however, as long as the set context is clear, the shorter name is sufficient.

A filter base (or filter basis) is a subset B of P(S) with the following properties:

1. The intersection of any two sets of B contains a set of B
2. B is non-empty and the empty set is not in B

Given a filter base B, one may obtain a (proper) filter by including all sets of P(S) which contain a set of B. The resulting filter is said to be generated by or spanned by filter base B. Every filter is also a filter base, so the process of passing from filter base to filter may be viewed as a sort of completion. If B and C are two filter bases on S, one says C is finer than B

(or that C is a refinement of B) if for each $B_0 \in B$, there is a $C_0 \in C$ such that $C_0 \subseteq B_0$.

- For filter bases B and C, if B is finer than C and C is finer than B, then B and C are said to be equivalent filter bases. Two filter bases are equivalent if and only if the filters they generate are equal.
- For filter bases A, B, and C, if A is finer than B and B is finer than C then A is finer than C. Thus the refinement relation is a preorder on the set of filter bases, and the passage from filter base to filter is an instance of passing from a preordering to the associated partial ordering.

Given a subset T of P(S) we can ask whether there exists a smallest filter F containing T. Such a filter exists if and only if the finite intersection of subsets of T is non-empty. We call T a subbase of F and say F is generated by T. F can be constructed by taking all finite intersections of T which is then filter base for F.

Examples

- Let S be a nonempty set and C be a nonempty subset. Then $\{C\}$ is a filter base. The filter it generates (i.e., the collection of all subsets containing C) is called the principal filter generated by C.
- A filter is said to be a free filter if the intersection of all of its members is empty. A principal filter is not free. Since the intersection of any finite number of members of a filter is also a member, no filter on a finite set is free, and indeed is the principal filter generated by the common intersection of all of its members. A nonprincipal filter on an infinite set is not necessarily free.
- The Fréchet filter on an infinite set S is the set of all subsets of S that have finite complement. The Fréchet filter is free, and it is contained in every free filter on S.
- Every uniform structure on a set X is a filter on $X \times X$.
- A filter in a poset can be created using the Rasiowa-Sikorski lemma, often used in forcing.

- The set $\{\{N, N+1, N+2, \ldots\} : N \in \{1,2,3,\ldots\}\}$ is called a *filter base of tails* of the sequence of natural numbers $(1,2,3,\ldots)$. A filter base of tails can be made of any net $(x_\alpha)_{\alpha \in A}$ using the construction $\{\{x_\alpha : \alpha \in A, \alpha_0 \leq a\} : \alpha_0 \in A\}$. Therefore, all nets generate a filter base (and therefore a filter). Since all sequences are nets, this holds for sequences as well.

Filters in Model Theory

For any filter *F* on a set *S*, the set function defined by

$$m(A) = \begin{cases} 1 & \text{if } A \in F \\ 0 & \text{if } S \setminus A \in F \\ \text{undefined} & \text{otherwise} \end{cases}$$

is finitely additive — a "measure" if that term is construed rather loosely. Therefore the statement

$$\{x \in S : \varphi(x)\} \in F$$

can be considered somewhat analogous to the statement that φ holds "almost everywhere". That interpretation of membership in a filter is used (for motivation, although it is not needed for actual *proofs*) in the theory of ultraproducts in model theory, a branch of mathematical logic.

Filters in Topology

In topology and analysis, filters are used to define convergence in a manner similar to the role of sequences in a metric space.

In topology and related areas of mathematics, a filter is a generalization of a net. Both nets and filters provide very general contexts to unify the various notions of limit to arbitrary topological spaces.

A sequence is usually indexed by the natural numbers, which are a totally ordered set. Thus, limits in first-countable spaces can be described by sequences. However, if the space is not first-countable, nets or filters must be used. Nets generalize the notion of a sequence by requiring the index set simply be a directed set. Filters can be thought of as sets built from multiple nets. Therefore, both the limit of a filter and the limit of a net are conceptually the same as the limit of a sequence.

Cauchy Filters

Let (X,d) be a metric space.

- To say that a filter base B on X is Cauchy means that for each real number $\varepsilon>0$, there is a $B_0 \in B$ such that the metric diameter of B_0 is less than ε.
- Take (x_n) to be a sequence in metric space X. (x_n) is a Cauchy sequence if and only if the filter base $\{\{x_N, x_{N+1}, ...\} : N \in \{1,2,3,...\}\}$ is Cauchy.

More generally, given a uniform space X, a filter F on X is called Cauchy filter if for every entourage U there is an $A \in F$ with $(x,y) \in U$ for all $x,y \in A$. In a metric space this agrees with the previous definition. X is said to be complete if every Cauchy filter converges. Conversely, on a uniform space every convergent filter is a Cauchy filter. Moreover, every cluster point of a Cauchy filter is a limit point.

A compact uniform space is complete: on a compact space each filter has a cluster point, and if the filter is Cauchy, such a cluster point is a limit point. Further, a uniformity is compact if and only if it is complete and totally bounded.

Most generally, a Cauchy space is a set equipped with a class of filters declared to be Cauchy. These are required to have the following properties:

1. for each x in X, the ultrafilter at x, $U(x)$, is Cauchy.
2. if F is a Cauchy filter, and F is a subset of a filter G, then G is Cauchy.
3. if F and G are Cauchy filters and each member of F intersects each member of G, then $F \cap G$ is Cauchy.

The Cauchy filters on a uniform space have these properties, so every uniform space (hence every metric space) defines a Cauchy space.

Chapter 3

Topological Space

There are many other equivalent ways to define a topological space: in other words, the concepts of neighbourhood or of open set can be reconstructed from the other starting points and satisfy the correct axioms. For example, using de Morgan's laws, the above axioms defining open sets become axioms defining closed sets:

1. The empty set and X are closed.
2. The intersection of any collection of closed sets is also closed.
3. The union of any pair of closed sets is also closed.

Using these axioms, another way to define a topological space is as a set X together with a collection of subsets of X satisfying the following axioms:

1. The empty set and X are in τ .
2. The intersection of any collection of sets in τ is also in τ .
3. The union of any pair of sets in τ is also in τ .

Under this definition, the sets in the topology τ are the closed sets, and their complements in X are the open sets.

Another way to define a topological space is by using the Kuratowski closure axioms, which define the closed sets as the fixed points of an operator on the power set of X. A net is a generalisation of the concept of sequence. A topology is completely determined if for every net in X the set of its accumulation points is specified. A variety of axiomatisations of topological spaces are listed in the Exercises of the book by

Vaidyanathaswamy. Thus one chooses the axiomatisation suited for the application, or the audience, as needed.

Comparison of Topologies

A variety of topologies can be placed on a set to form a topological space. When every set in a topology τ_1 is also in a topology τ_2, we say that τ_2 is *finer* than τ_2, and τ_1 is *coarser* than τ_2. A proof that relies only on the existence of certain open sets will also hold for any finer topology, and similarly a proof that relies only on certain sets not being open applies to any coarser topology.

The terms *larger* and *smaller* are sometimes used in place of finer and coarser, respectively. The terms *stronger* and *weaker* are also used in the literature, but with little agreement on the meaning, so one should always be sure of an author's convention when reading. The collection of all topologies on a given fixed set *X* forms a complete lattice: if $F = \{\ \tau_\alpha$: αin A} is a collection of topologies on *X*, then the meet of *F* is the intersection of *F*, and the join of *F* is the meet of the collection of all topologies on *X* that contain every member of *F*.

Continuous Functions

A function $f : X \to Y$ between topological spaces is called continuous if for all $x \in X$ and all neighbourhoods *N* of $f(x)$ there is a neighbourhood *M* of *x* such that $f(M) \subseteq N$. This relates easily to the usual definition in analysis. Equivalently, *f* is continuous if the inverse image of every open set is open. This is an attempt to capture the intuition that there are no "jumps" or "separations" in the function. A homeomorphism is a bijection that is continuous and whose inverse is also continuous.

Two spaces are called homeomorphic if there exists a homeomorphism between them. From the standpoint of topology, homeomorphic spaces are essentially identical. In category theory, Top, the category of topological spaces with topological spaces as objects and continuous functions as morphisms is one of the fundamental categories in mathematics. The attempt to classify the objects of this category (up to homeomorphism) by invariants has motivated and generated entire areas of research, such as homotopy theory, homology theory, and K-theory, to name just a few.

Examples of Topological Spaces

A given set may have many different topologies. If a set is given a different topology, it is viewed as a different topological space. Any set can be given the discrete topology in which every subset is open. The only convergent sequences or nets in this topology are those that are eventually constant. Also, any set can be given the trivial topology (also called the indiscrete topology), in which only the empty set and the whole space are open. Every sequence and net in this topology converges to every point of the space. This example shows that in general topological spaces, limits of sequences need not be unique. However, often topological spaces must be Hausdorff spaces where limit points are unique.

There are many ways of defining a topology on R, the set of real numbers. The standard topology on R is generated by the open intervals. The set of all open intervals forms a base or basis for the topology, meaning that every open set is a union of some collection of sets from the base. In particular, this means that a set is open if there exists an open interval of non zero radius about every point in the set. More generally, the Euclidean spaces R^n can be given a topology.

In the usual topology on R^n the basic open sets are the open balls. Similarly, C and C^n have a standard topology in which the basic open sets are open balls. Every metric space can be given a metric topology, in which the basic open sets are open balls defined by the metric. This is the standard topology on any normed vector space.

On a finite-dimensional vector space this topology is the same for all norms. Many sets of linear operators in functional analysis are endowed with topologies that are defined by specifying when a particular sequence of functions converges to the zero function. Any local field has a topology native to it, and this can be extended to vector spaces over that field. Every manifold has a natural topology since it is locally Euclidean. Similarly, every simplex and every simplicial complex inherits a natural topology from R^n.

The Zariski topology is defined algebraically on the spectrum of a ring or an algebraic variety. On R^n or C^n, the closed sets of the Zariski topology are the solution sets of systems of polynomial equations. A linear graph has a natural topology that generalises many of the geometric aspects of graphs with vertices and edges.

The Sierpiñski space is the simplest non-discrete topological space. It has important relations to the theory of computation and semantics.

There exist numerous topologies on any given finite set. Such spaces are called finite topological spaces. Finite spaces are sometimes used to provide examples or counterexamples to conjectures about topological spaces in general. Any set can be given the cofinite topology in which the open sets are the empty set and the sets whose complement is finite. This is the smallest T_1 topology on any infinite set.

Any set can be given the cocountable topology, in which a set is defined as open if it is either empty or its complement is countable. When the set is uncountable, this topology serves as a counterexample in many situations. The real line can also be given the lower limit topology. Here, the basic open sets are the half open intervals $[a, b)$. This topology on R is strictly finer than the Euclidean topology defined above; a sequence converges to a point in this topology if and only if it converges from above in the Euclidean topology. This example shows that a set may have many distinct topologies defined on it.

If Γ is an ordinal number, then the set $\Gamma = [0, \Gamma)$ may be endowed with the order topology generated by the intervals (a, b), $[0, b)$ and (a, Γ) where a and b are elements of Γ.

Topological Constructions

Every subset of a topological space can be given the subspace topology in which the open sets are the intersections of the open sets of the larger space with the subset. For any indexed family of topological spaces, the product can be given the product topology, which is generated by the inverse images of open sets of the factors under the projection mappings. For example, in finite products, a basis for the product topology consists of all products of open sets. For infinite products, there is the additional requirement that in a basic open set, all but finitely many of its projections are the entire space.

A quotient space is defined as follows: if X is a topological space and Y is a set, and if $f : X \to Y$ is a surjective function, then the quotient topology on Y is the collection of subsets of Y that have open inverse images under f. In other words, the quotient topology is the finest topology on Y for which f is continuous. A common example of a quotient topology is when an equivalence relation is defined on the topological space X. The map f is then the natural projection onto the set of equivalence classes. The Vietoris topology on the set of all non-empty subsets of a topological space X, named for Leopold Vietoris, is generated by the following basis: for every n-tuple $U_1, \ldots, U_n$ of open sets in X, we construct a basis set consisting of all subsets of the union of the U_i that have non-empty intersections with each U_i.

Classification of Topological Spaces

Topological spaces can be broadly classified, up to homeomorphism, by their topological properties. A topological property is a property of spaces that is invariant under homeomorphisms. To prove that two spaces are not homeomorphic it is sufficient to find a topological property not shared by them. Examples of such properties include connectedness, compactness, and various separation axioms.

Topological Spaces with Algebraic Structure

For any algebraic objects we can introduce the discrete topology, under which the algebraic operations are continuous functions. For any such structure that is not finite, we often have a natural topology compatible with the algebraic operations, in the sense that the algebraic operations are still continuous. This leads to concepts such as topological groups, topological vector spaces, topological rings and local fields.

Topological Spaces with Order Structure

- Spectral. A space is spectral if and only if it is the prime spectrum of a ring (Hochster theorem).
- Specialisation preorder. In a space the specialisation (or canonical) preorder is defined by $x \leq y$ if and only if $\mathrm{cl}\{x\} \subseteq \mathrm{cl}\{y\}$.

Homology Theory

In mathematics, homology theory is the axiomatic study of the intuitive geometric idea of *homology of cycles* on topological spaces. It can be broadly defined as the study of homology theories on topological spaces.

General Idea

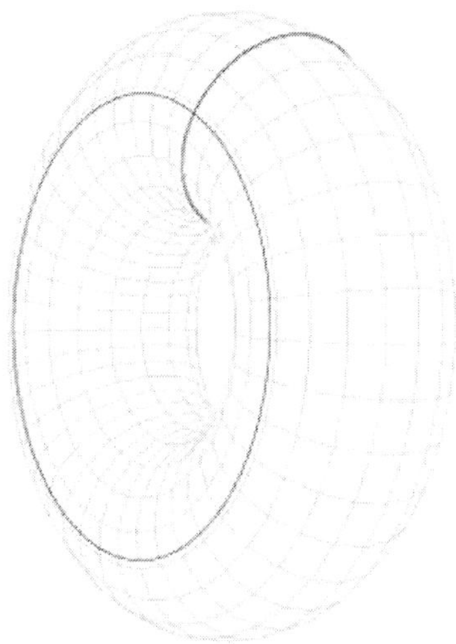

Figure: *A torus with generators coloured in pink and red.*

To any topological space X and any natural number k, one can associate a set $H_k(X)$, whose elements are called (k-dimensional) homology classes. There is a well-defined way to add and subtract homology classes, which makes $H_k(X)$ into an abelian group, called the kth homology group of X. In heuristic terms, the size and structure of $H_k(X)$ gives information about the number of k-dimensional holes in X.

For example, if $X$is a figure eight, then it has two holes, which in this context count as being one-dimensional. The corresponding homology group $H_1(X)$ can be identified with the group $\mathbb{Z}\oplus\mathbb{Z}$ of pairs of integers, with one copy of $\mathbb{Z}$ for each hole. While it seems very straightforward to say that X has two holes, it is surprisingly hard to formulate this in a mathematically rigorous way; this is a central purpose of homology theory.

For a more intricate example, if Y is a Klein bottle then $H_1(Y)$ can be identified with $\mathbb{Z} \oplus \mathbb{Z}/2\mathbb{Z}$. This is not just a sum of copies of $\mathbb{Z}$, so it gives more subtle information than just a count of holes.

The formal definition of $H_1(X)$ can be sketched as follows. The elements $H_1(X)$ of are one-dimensional cycles, except that two cycles are considered to represent the same element if they are homologous. The simplest kind of one-dimensional cycles are just closed curves in , which could consist of one or more loops. If a closed curve C_0 can be deformed continuously within $X$to another closed curve C_1, then $C_0$and $C_1$are homologous and so determine the same element of $H_1(X)$. This captures the main geometric idea, but the full definition is somewhat more complex. There is also a version (called simplicial homology) that works when X is presented as a simplicial complex; this is smaller and easier to understand, but technically less flexible.

For example, let T be a torus, as shown on the right. Let C be the pink curve, and let D be the red one. For integers $n$and m, we have another closed curve that goes n times around C and then m times around D; this is denoted by $nC + mD$. It can be shown that any closed curve in $T$is homologous to $nC + mD$ for some n and m, and thus that $H_1(T)$ is again isomorphic to $\mathbb{Z} \oplus \mathbb{Z}$.

Cohomology

As well as the homology groups $H_k(X)$, one can define cohomology groups $H^k(X)$. In the common case where each group $H_k(X)$ is isomorphic to $\mathbb{Z}^{r_k}$ for some $r_k \in \mathbb{N}$, we just have $H^k(X) = Hom(H_k(X), \mathbb{Z})$, which is again isomorphic to $\mathbb{Z}^{r_k}$, and $H_k(X) = Hom(H^k(X), \mathbb{Z})$, so $H_k(X)$ and $H^k(X)$ determine each other. In general, the relationship between $H_k(X)$and $H^k(X)$ is only a little more complicated, and is controlled by the universal coefficient theorem. The main advantage of cohomology over homology is that it has a natural ring structure: there is a way to multiply an i-dimensional cohomology class by a j-dimensional cohomology class to get an $i + j$ -dimensional cohomology class.

Applications

Notable theorems proved using homology include the following:

- *The Brouwer fixed point theorem:* If *f* is any continuous map from the ball B^n to itself, then there is a fixed point $a \in B^n$ with $f(a) = a$.
- *Invariance of domain:* If *U* is an open subset of $\mathbb{R}^n$ and $f : U \to \mathbb{R}^n$ is an injective continuous map, then $V = f(U)$ is open and *f* is a homeomorphism between *U* and *V*.
- *The Hairy ball theorem:* any vector field on the $2k$-sphere (or more generally, the -sphere for any $k \geq 1$) vanishes at some point.
- *The Borsuk–Ulam theorem:* any continuous function from an *n*-sphere into Euclidean *n*-space maps some pair of antipodal points to the same point. (Two points on a sphere are called antipodal if they are in exactly opposite directions from the sphere's centre.)

Intersection Theory and Poincaré Duality

Let *M*be a compact oriented manifold of dimension *n*. The Poincaré duality theorem gives a natural isomorphism $H^k(M) \simeq H_{n-k}(M)$, which we can use to transfer the ring structure from cohomology to homology. For any compact oriented submanifold $N \subseteq M$ of dimension *d*, one can define a so-called fundamental class $[N] \in H_d(M) \simeq H^{n-d}(M)$. If *L*is another compact oriented submanifold which meets *N* transversely, it works out that $[L][N] = [L \cap N]$. In many cases the group $H_d(M)$ will have a basis consisting of fundamental classes of submanifolds, in which case the product rule $[L][N] = [L \cap N]$ gives a very clear geometric picture of the ring structure.

Connection with Integration

Suppose that *X*is an open subset of the complex plane, that $f(z)$ is a holomorphic function on *X*, and that *C* is a closed curve in *X*. There is then a standard way to define the contour integral $\oint_C f(z)dz$, which is a central idea in complex analysis.

One formulation of Cauchy's integral theorem is as follows: if C_0 and C_1 are homologous, then $\oint_{C_0} f(z)dz = \oint_{C_1} f(z)dz$. (Many authors consider only the case where *X* is simply connected, in which case every closed curve is homologous to the empty curve and so $\oint_C f(z)dz = 0$.) This means that we can make sense of $\oint_c f(z)dz$ when is merely a homology class, or in other words an element of $H_1(X)$. It is also important that in the case where $f(z)$ is the derivative of another function $g(z)$, we always have $\oint_C g'(z)dz = 0$ (even when C is not homologous to zero).

This is the simplest case of a much more general relationship between homology and integration, which is most efficiently formulated in terms of differential forms and de Rham cohomology. To explain this briefly, suppose that *X* is an open subset of $\mathbb{R}^N$, or more generally, that *X* is a manifold. One can then define objects called *n*-forms on . *X* If *X* is open in $\mathbb{R}^3$, then the 0-forms are just the scalar fields, the 1-forms are the vector fields, the 2-forms are the same as the 1-forms, and the 3-forms are the same as the 0-forms.

There is also a kind of differentiation operation called the exterior derivative: if is an -form, then the exterior derivative is an $(n+1)$-form denoted by $d\alpha$. The standard operators div, grad and curl from vector calculus can be seen as special cases of this. There is a procedure for integrating an *n*-form α over an *n*-cycle C to get a number $\oint_C \alpha$. It can be shown that $\oint_C d\beta = 0$ for any $(n-1)$-form β, and that $\oint_C \alpha$ depends only on the homology class of *C*, provided that $d\alpha = 0$. The classical Stokes's Theorem and Divergence Theorem can be seen as special cases of this.

We say that α is *closed* if $d\alpha = 0$, and *exact* if $\alpha = d\beta$ for some β. It can be shown that $dd\beta$ is always zero, so that exact forms are always closed. The de Rham cohomology group $H^k_{dR}(X)$ is the quotient of the group of closed forms by the subgroup of exact forms. It follows from the above that there is a well-defined pairing $H_k(X) \times H^k_{dR}(X) \to \mathbb{R}$ given by integration.

Axiomatics and Generalised Homology

There are various ways to define cohomology groups (for example singular cohomology, Čech cohomology, Alexander–Spanier cohomology or Sheaf cohomology). These give different answers for some exotic spaces, but there is a large class of spaces on which they all agree. This is most easily understood axiomatically: there is a list of properties known as the Eilenberg–Steenrod axioms, and any two constructions that share those properties will agree at least on all finite CW complexes, for example.

One of the axioms is the so-called dimension axiom: if P is a single point, then $H_n(P) = 0$ for all $n \neq 0$, and $H_0(P) = \mathbb{Z}$. We can generalise slightly by allowing an arbitrary abelian group A in dimension zero, but still insisting that the groups in nonzero dimension are trivial. It turns out that there is again an essentially unique system of groups satisfying these axioms, which are denoted by $H_*(X; A)$. In the common case where each group $H_k(X)$ is isomorphic to $\mathbb{Z}^{r_k}$ for some $r_k \in \mathbb{N}$, we just have $H_k(X; A) = A^{r_k}$. In general, the relationship between $H_k(X)$ and $H_k(X; A)$ is only a little more complicated, and is again controlled by the Universal coefficient theorem.

More significantly, we can drop the dimension axiom altogether. There are a number of different ways to define groups satisfying all the other axioms, including the following:

- The stable homotopy groups $\pi_k^S(X)$
- Various different flavours of cobordism groups: $MO_*(X)$, $MSO_*(X)$, $MU_*(X)$ and so on. The last of these (known as complex cobordism) is especially important, because of the link with formal group theory via a theorem of Daniel Quillen.
- Various different flavours of K-theory: $KO_*(X)$ (real periodic K-theory), $kO_*(X)$ (real connective), $KU_*(X)$ (complex periodic), $kU_*(X)$ (complex connective) and so on.
- Brown–Peterson homology, Morava K-theory, Morava E-theory, and other theories defined using the algebra of formal groups.

- Various flavours of elliptic homology

These are called generalised homology theories; they carry much richer information than ordinary homology, but are often harder to compute. Their study is tightly linked (via the Brown representability theorem) to stable homotopy.

Homological Algebra and Homology of Other Objects

A chain complex consists of groups C_i (for all $i \in \mathbb{Z}$) and homomorphisms $d: C_i \to C_{i-1}$ satisfying $dd = 0$. This condition shows that the groups $B_i = \text{image}(d: C_{i+1} \to C_i)$ are contained in the groups $Z_i = \ker(d: C_i \to C_{i-1})$, so one can form the quotient groups $H_i = Z_i / B_i$, which are called the homology groups of the original complex. There is a similar theory of cochain complexes, consisting of groups C^i and homomorphisms $\delta: C^i \to C^{i+1}$.

The simplicial, singular, Èech and Alexander–Spanier groups are all defined by first constructing a chain complex or cochain complex, and then taking its homology. Thus, a substantial part of the work in setting up these groups involves the general theory of chain and cochain complexes, which is known as homological algebra. One can also associate (co)chain complexes to a wide variety of other mathematical objects, and then take their (co)homology. For example, there are cohomology modules for groups, Lie algebras and so on.

Cohomology

In mathematics, specifically in algebraic topology, cohomology is a general term for a sequence of abelian groups defined from a co-chain complex.

That is, cohomology is defined as the abstract study of cochains, cocycles, and coboundaries. Cohomology can be viewed as a method of assigning algebraic invariants to a topological space that has a more refined algebraic structure than does homology. Cohomology arises from the algebraic dualization of the construction of homology. In less abstract language, cochains in the fundamental sense should assign 'quantities' to the *chains* of homology theory.

From its beginning in topology, this idea became a dominant method in the mathematics of the second half of the twentieth century; from the initial idea of *homology* as a topologically invariant relation on *chains*, the range of applications of homology and cohomology theories has spread out over geometry and abstract algebra.

The terminology tends to mask the fact that in many applications *cohomology*, a contravariant theory, is more natural than *homology*.

At a basic level this has to do with functions and pullbacks in geometric situations: given spaces X and Y, and some kind of function F on Y, for any mapping $f: X \to Y$ composition with f gives rise to a function $F \circ f$ on X. Cohomology groups often also have a natural product, the cup product, which gives them a ring structure. Because of this feature, cohomology is a stronger invariant than homology, as it can differentiate between certain algebraic objects that homology cannot.

Definition

For a topological space X, the cohomology group $H^n(X;G)$, with coefficents in G, is defined to be the quotient $\mathrm{Ker}(\delta^n)/\mathrm{Im}(\delta^{n-1})$ at $C^n(X;G)$ in the cochain complex

$$\cdots \leftarrow C^{n+1}(X;G) \overset{\delta^n}{\leftarrow} C^n(X;G) \overset{\delta^{n-1}}{\leftarrow} C^{n-1}(X;G) \leftarrow \cdots \leftarrow C^0(X;G) \leftarrow 0.$$

Elements in $\mathrm{Ker}(\delta^n)$ are n-cocycles and elements in $\mathrm{Im}(\delta^{n-1})$ are n-coboundaries. The cohomology groups with $n \geq 1$ are called the *higher cohomology*.

History

Although cohomology is fundamental to modern algebraic topology, its importance was not seen for some 40 years after the development of homology. The concept of *dual cell structure*, which Henri Poincaré used in his proof of his Poincaré duality theorem, contained the germ of the idea of cohomology, but this was not seen until later.

There were various precursors to cohomology. In the mid-1920s, J.W. Alexander and Solomon Lefschetz founded the intersection theory of cycles on manifolds. On an n-dimensional manifold M, a p-cycle and a q-cycle with nonempty intersection will, if in general position, have intersection a $(p + q - n)$-cycle. This enables us to define a multiplication of homology classes

$$H_p(M) \times H_q(M) \to H_{p+q-n}(M).$$

Alexander had by 1930 defined a first cochain notion, based on a p-cochain on a space X having relevance to the small neighbourhoods of the diagonal in X^{p+1}.

In 1931, Georges de Rham related homology and exterior differential forms, proving De Rham's theorem. This result is now understood to be more naturally interpreted in terms of cohomology.

In 1934, Lev Pontryagin proved the Pontryagin duality theorem; a result on topological groups. This (in rather special cases) provided an interpretation of Poincaré duality and Alexander duality in terms of group characters. At a 1935 conference in Moscow, Andrey Kolmogorov and Alexander both introduced cohomology and tried to construct a cohomology product structure.

In 1936 Norman Steenrod published a paper constructing Èech cohomology by dualizing Èech homology. From 1936 to 1938, Hassler Whitney and Eduard Èech developed the cup product (making cohomology into a graded ring) and cap product, and realised that Poincaré duality can be stated in terms of the cap product. Their theory was still limited to finite cell complexes.

In 1944, Samuel Eilenberg overcame the technical limitations, and gave the modern definition of singular homology and cohomology.

In 1945, Eilenberg and Steenrod stated the axioms defining a homology or cohomology theory. In their 1952 book, *Foundations of Algebraic Topology*, they proved that the existing homology and cohomology theories did indeed satisfy their axioms.

In 1948 Edwin Spanier, building on work of Alexander and Kolmogorov, developed Alexander–Spanier cohomology.

Cohomology Theories

Eilenberg–Steenrod Theories: A *cohomology theory* is a family of contravariant functors from the category of pairs of topological spaces and continuous functions (or some subcategory thereof such as the category of CW complexes) to the category of Abelian groups and group homomorphisms that satisfies the Eilenberg–Steenrod axioms.

Some cohomology theories in this sense are:

- simplicial cohomology
- singular cohomology
- de Rham cohomology
- Èech cohomology
- sheaf cohomology.

Generalized Cohomology Theories

When one axiom (the *dimension axiom*) is relaxed, one obtains the idea of generalized cohomology theory or extraordinary cohomology theory; this allows theories based on K-theory and cobordism theory. There are others, coming from stable homotopy theory. In this context, singular homology is referred to as ordinary homology.

A generalized cohomology theory is "determined by its values on a point", in the sense that if one has a space given by contractible spaces (homotopy equivalent to a point), glued together along contractible spaces, as in a simplicial complex, then the cohomology of the space is determined by the cohomology of a point and the combinatorics of the patching, and effectively computable. Formally, this is computed by the excision theorem, or equivalently the Mayer–Vietoris sequence. Thus the cohomology of a point is a fundamental calculation for any generalized cohomology theory, though the cohomology of particular spaces is also of interest.

One reason that generalized cohomology theories are interesting is that they are representable functors if one works in a larger category than CW complexes; namely, the category of spectra.

Other Cohomology Theories

Theories in a broader sense of *cohomology* include:

- André–Quillen cohomology
- BRST cohomology
- Bonar–Claven cohomology
- Bounded cohomology
- Coherent cohomology

- Crystalline cohomology
- Cyclic cohomology
- Deligne cohomology
- Dirac cohomology
- Étale cohomology
- Flat cohomology
- Galois cohomology
- Gel'fand–Fuks cohomology
- Group cohomology
- Harrison cohomology
- Hochschild cohomology
- Intersection cohomology
- Lie algebra cohomology
- Local cohomology
- Motivic cohomology
- Non-abelian cohomology
- Perverse cohomology
- Quantum cohomology
- Schur cohomology
- Spencer cohomology
- Topological André–Quillen cohomology
- Topological cyclic cohomology
- Topological Hochschild cohomology
- G cohomology

List of Cohomology Theories

This is a list of some of the ordinary and generalized (or extraordinary) homology and cohomology theories in algebraic topology that are defined on the categories of CW complexes or spectra. For other sorts of homology theories see the links at the end of this article.

Notation

- S = p = S^0 is the sphere spectrum.
- S^n is the spectrum of the n-dimensional sphere

- $S^nY = S^{n'}\text{-}Y$ is the nth suspension of a spectrum Y.
- $[X, Y]$ is the abelian group of morphisms from the spectrum X to the spectrum Y, given (roughly) as homotopy classes of maps.
- $[X, Y]_n = [S^nX, Y]$
- $[X, Y]_*$ is the graded abelian group given as the sum of the groups $[X, Y]_n$.
- $p_n(X) = [S^n, X] = [S, X]_n$ is the nth stable homotopy group of X.
- $p_*(X)$ is the sum of the groups $p_n(X)$, and is called the coefficient ring of X when X is a ring spectrum.
- X"Y is the smash product of two spectra.

If X is a spectrum, then it defines generalized homology and cohomology theories on the category of spectra as follows.

- $X_n(Y) = [S, X \neq Y]_n = [S^n, X \neq Y]$ is the generalized homology of Y,
- $X^n(Y) = [Y, X]_{"n} = [S^{"n}Y, X]$ is the generalized cohomology of Y

Ordinary Homology Theories

These are the theories satisfying the "dimension axiom" of the Eilenberg–Steenrod axioms that the homology of a point vanishes in dimension other than 0.

They are determined by an abelian coefficient group G, and denoted by H(X, G) (where G is sometimes omitted, especially if it is Z).

Usually G is the integers, the rationals, the reals, the complex numbers, or the integers mod a prime p.

The cohomology functors of ordinary cohomology theories are represented by Eilenberg–MacLane spaces. On simplicial complexes, these theories coincide with singular homology and cohomology.

Homology and Cohomology with Integer Coefficients:

Spectrum: H (Eilenberg–MacLane spectrum of the integers.)

Coefficient Ring: $\pi_n(\text{H}) = \text{Z}$ if $n = 0$, 0 otherwise.

The original homology theory.

Homology and Cohomology with Rational (or Real or Complex) Coefficients:

Spectrum: HQ (Eilenberg–Mac Lane spectrum of the rationals.)

Coefficient Ring: $\pi_n(\mathrm{HQ}) = \mathrm{Q}$ if $n = 0$, 0 otherwise.

These are the easiest of all homology theories. The homology groups $\mathrm{HQ}_n(X)$ are often denoted by $\mathrm{H}_n(X, Q)$. The homology groups H(*X*, Q), H(*X*, R), H(*X*, C) with rational, real, and complex coefficients are all similar, and are used mainly when torsion is not of interest (or too complicated to work out).

The Hodge decomposition writes the complex cohomology of a complex projective variety as a sum of sheaf cohomology groups.

Homology and Cohomology with Mod p Coefficients:

Spectrum: HZ_p (Eilenberg–Maclane spectrum of the integers mod *p*.)

Coefficient Ring: $\pi_n(\mathrm{HZ}_p) = \mathrm{Z}_p$ (Integers mod *p*) if $n = 0$, 0 otherwise.

K-Theories

The simpler K-theories of a space are often related to vector bundles over the space, and different sorts of K-theories correspond to different structures that can be put on a vector bundle.

Real K-Theory

Spectrum: KO

Coefficient Ring: The coefficient groups $\pi_i(\mathrm{KO})$ have period 8 in *i*, given by the sequence Z, Z_2, Z_2,0, Z, 0, 0, 0, repeated. As a ring, it is generated by a class η in degree 1, a class x_4 in degree 4, and an invertible class v_1^4 in degree 8, subject to the relations that $2\eta = \eta^3 = \eta x_4 = 0$, and $x_4^2 = 4v_1^4$.

$\mathrm{KO}^0(X)$ is the ring of stable equivalence classes of real vector bundles over *X*. Bott periodicity implies that the K-groups have period 8.

Complex K-Theory

Spectrum: KU (even terms BU or Z × BU, odd terms *U*).

Coefficient Ring: The coefficient ring K^*(point) is the ring of formal Laurent series in a generator of degree 2.

$K^0(X)$ is the ring of stable equivalence classes of complex vector bundles over X. Bott periodicity implies that the K-groups have period 2.

Quaternionic K-Theory

Spectrum: KSp

Coefficient Ring: The coefficient groups π_i(KSp) have period 8 in i, given by the sequence Z, 0, 0, 0,Z, Z_2, Z_2,0, repeated.

$KSp^0(X)$ is the ring of stable equivalence classes of quaternionic vector bundles over X. Bott periodicity implies that the K-groups have period 8.

K Theory with Coefficients

Spectrum: KG

G is some abelian group; for example the localization $Z_{(p)}$ at the prime p. Other K-theories can also be given coefficients.

Self Conjugate K-Theory

Spectrum: KSC

Coefficient Ring: *to be written...*

The coefficient groups π_i(KSC) have period 4 in i, given by the sequence Z, Z_2, 0, Z, repeated. Introduced by D. Anderson in his unpublished 1964 Berkeley PhD dissertation, "A new cohomology theory".

Connective K-Theories

Spectrum: ku for connective K-theory, ko for connective real K-theory.

Coefficient Ring: For ku, the coefficient ring is the ring of polynomials over Z on a single class v_1 in dimension 2.

For ko, the coefficient ring is the quotient of a polynomial ring on three generators, η in dimension 1, x_4 in dimension 4, and v_1^4 in dimension 8, the periodicity generator, modulo the relations that $2\eta = 0$, $x_4^2 = 4v_1^4$, $\eta^3 = 0$, and $\eta x = 0$. Roughly speaking, this is K-theory with the negative dimensional parts killed off.

KR-Theory

This is a cohomology theory defined for spaces with involution, from which many of the other K-theories can be derived.

Bordism and Cobordism Theories

Cobordism studies manifolds, where a manifold is regarded as "trivial" if it is the boundary of another compact manifold. The cobordism classes of manifolds form a ring that is usually the coefficient ring of some generalized cohomology theory. There are many such theories, corresponding roughly to the different structures that one can put on a manifold. The functors of cobordism theories are often represented by Thom spaces of certain groups.

Stable Homotopy and Cohomotopy

Spectrum: S (sphere spectrum).

Coefficient ring: The coefficient groups $\pi_n(S)$ are the stable homotopy groups of spheres, which are notoriously hard to compute or understand for $n > 0$. (For $n < 0$ they vanish, and for $n = 0$ the group is Z.) Stable homotopy is closely related to cobordism of framed manifolds (manifolds with a trivialization of the normal bundle).

Unoriented Cobordism

Spectrum: MO (Thom spectrum of orthogonal group)

Coefficient Ring: π_*(MO) is the ring of cobordism classes of unoriented manifolds, and is a polynomial ring over the field with 2 elements on generators of degree i for every i not of the form 2^{n}"1.

Unoriented bordism is 2-torsion, since *2M* is the boundary of.

MO is a rather weak cobordism theory, as the spectrum MO is isomorphic to H(π_*(MO)) ("homology with coefficients in π_*(MO)") – MO is a product of Eilenberg–MacLane spectra.

In other words the corresponding homology and cohomology theories are no more powerful than homology and cohomology with coefficients in Z/2Z. This was the first cobordism theory to be described completely.

Complex Cobordism

Spectrum: MU (Thom spectrum of unitary group)

Coefficient Ring: $\pi_*(MU)$ is the polynomial ring on generators of degree 2, 4, 6, 8, ... and is naturally isomorphic to Lazard's universal ring, and is the cobordism ring of stably almost complex manifolds.

Oriented Cobordism

Spectrum: MSO (Thom spectrum of special orthogonal group)

Coefficient ring: The oriented cobordism class of a manifold is completely determined by its characteristic numbers: its Stiefel–Whitney numbers and Pontryagin numbers, but the overall coefficient ring, denoted $\Omega_* = \Omega(*) = MSO(*)$ is quite complicated.

Rationally, and at 2 (corresponding to Pontryagin and Stiefel–Whitney classes, respectively), MSO is a product of Eilenberg–MacLane spectra – $MSO_Q = H(\pi_*(MSO_Q))$ and $MSO[2] = H(\pi_*(MSO[2]))$ – but at odd primes it is not, and the structure is complicated to describe. The ring has been completely described integrally, due to work of Milnor, Averbuch, Rokhlin, and C. T. C. Wall.

Special Unitary Cobordism

Spectrum: MSU (Thom spectrum of special unitary group)

Coefficient Ring:

Spin Cobordism (and Variants)

Spectrum: MSpin (Thom spectrum of spin group)

Coefficient Ring:

Symplectic Cobordism

Spectrum: MSp (Thom spectrum of symplectic group)

Coefficient Ring:

Clifford Algebra Cobordism

PL cobordism and topological cobordism

Spectrum: MPL, MSPL, MTop, MSTop

Coefficient Ring:

The definition is similar to cobordism, except that one uses piecewise linear or topological instead of smooth manifolds, either oriented or unoriented. The coefficient rings are complicated.

Brown–Peterson Cohomology

Spectrum: BP

Coefficient Ring: $\pi_*(\mathrm{BP})$ is a polynomial algebra over $Z_{(p)}$ on generators v_n of dimension $2(p^n - 1)$ for $n \geq 1$.

Brown–Peterson cohomology BP is a summand of MU_p, which is complex cobordism MU localized at a prime p. In fact $\mathrm{MU}_{(p)}$ is a sum of suspensions of BP.

Morava K-Theory

Spectrum: K(n) (They also depend on a prime p.)

Coefficient Ring: $\mathrm{F}_p[v_n, v_n^{"1}]$, where v_n has degree $2(p^n$ " 1).

These theories have period $2(p^n$ " 1). They are named after Jack Morava.

Johnson–Wilson Theory

Spectrum $E(n)$

Coefficient ring $\mathrm{Z}_{(2)}[v_1, ..., v_n, 1/v_n]$ where v_i has degree $2(2^{i"}1)$

String cobordism

Spectrum:

Coefficient ring:

Theories Related to Elliptic Curves

Elliptic cohomology:

Spectrum: Ell

Topological Modular Forms

Spectra: tmf, TMF (previously called eo_2.)

The coefficient ring $\pi_*(\mathrm{tmf})$ is called the ring of topological modular forms. TMF is tmf with the 24th power of the modular form Δ inverted, and has period 24^2=576. At the prime $p = 2$, the completion of tmf is the spectrum eo_2, and the K(2)-localization of tmf is the Hopkins-Miller Higher Real K-theory spectrum EO_2.

Combinatorial Topology

In mathematics, combinatorial topology was an older name for algebraic topology, dating from the time when topological invariants of spaces (for example the Betti numbers) were regarded as derived from combinatorial decompositions such as simplicial complexes.

After the proof of the simplicial approximation theorem this approach provided rigour. The change of name reflected the move to organise topological classes such as cycles modulo boundaries explicitly into abelian groups. This point of view is often attributed to Emmy Noether, and so the change of title may reflect her influence.

The transition is also attributed to the work of Heinz Hopf, who was influenced by Noether, and to Leopold Vietoris and Walther Mayer, who independently defined homology. A fairly precise date can be supplied in the internal notes of the Bourbaki group. While topology was still *combinatorial* in 1942, it had become *algebraic* by 1944.

Rosenfeld (1979) proposed digital topology for image processing that can be considered as a new development of combinatorial topology. The digital forms of Euler characteristic theorem and Gauss–Bonnet theorem were obtained by Chen et al. In history, a 2D grid cell topology was appeared in Alexandrov-Hopf's book Topologie I (1935).

Continuum (Topology)

In the mathematical field of point-set topology, a continuum (pl *continua*) is a nonempty compact connected metric space, or less frequently, a compact connected Hausdorff space. Continuum theory is the branch of topology devoted to the study of continua

Definitions

- A continuum that contains more than one point is called nondegenerate.
- A subset A of a continuum X such that A itself a continuum is called a subcontinuum of X. A space homeomorphic to a subcontinuum of the Euclidean plane R^2 is called a planar continuum.

- A continuum X is homogeneous if for every two points x and y in X, there exists a homeomorphism $h: X \to X$ such that $h(x) = y$.
- A Peano continuum is a continuum that is locally connected at each point.
- An indecomposable continuum is a continuum that cannot be represented as the union of two proper subcontinua. A continuum X is hereditarily indecomposable if every subcontinuum of X is indecomposable.
- The dimension of a continuum usually means its topological dimension. A one-dimensional continuum is often called a curve.

Examples

- An arc is a space homeomorphic to the closed interval [0,1]. If h: $[0,1] \to X$ is a homeomorphism and $h(0) = p$ and $h(1) = q$ then p and q are called the endpoints of X; one also says that X is an arc from p to q. An arc is the simplest and most familiar type of a continuum. It is one-dimensional, arcwise connected, and locally connected.
- Topologist's sine curve is a subset of the plane which is the union of the graph of the function $f(x) = \sin(1/x)$, $0 < x \leq 1$ with the segment “1 ≤ y ≤ 1 of the y-axis. It is a one-dimensional continuum that is not arcwise connected, and it is locally disconnected at the points along the y-axis.
- The Warsaw circle is obtained by “closing up” the topologist's sine curve by an arc connecting (0,”1) and (1,sin(1)). It is a one-dimensional continuum whose homotopy groups are all trivial, but it is not a contractible space.
- An n-cell is a space homeomorphic to the closed ball in the Euclidean space R^n. It is contractible and is the simplest example of an n-dimensional continuum.
- An n-sphere is a space homeomorphic to the standard n-sphere in the $(n + 1)$-dimensional Euclidean space. It is an n-dimensional homogeneous continuum that is not contractible, and therefore different from an n-cell.

- The Hilbert cube is an infinite-dimensional continuum.
- Solenoids are among the simplest examples of indecomposable homogeneous continua. They are neither arcwise connected nor locally connected.
- Sierpinski carpet, also known as the *Sierpinski universal curve*, is a one-dimensional planar Peano continuum that contains a homeomorphic image of any one-dimensional planar continuum.
- Pseudo-arc is a homogeneous hereditarily indecomposable planar continuum.

Properties

There are two fundamental techniques for constructing continua, by means of *nested intersections* and *inverse limits.*

- If $\{X_n\}$ is a nested family of continua, i.e. $X_n \supseteq X_{n+1}$, then their intersection is a continuum.
- If $\{(X_n, f_n)\}$ is an inverse sequence of continua X_n, called the coordinate spaces, together with continuous maps $f_n: X_{n+1} \to X_n$, called the bonding maps, then its inverse limit is a continuum.

A finite or countable product of continua is a continuum.

Uniform Continuity

Similar to continuous functions between topological spaces, which preserve topological properties, are the uniform continuous functions between uniform spaces, which preserve uniform properties. Uniform spaces with uniform maps form a category. An isomorphism between uniform spaces is called a uniform isomorphism. A uniformly continuous function is defined as one where inverse images of entourages are again entourages, or equivalently, one where the inverse images of uniform covers are again uniform covers. All uniformly continuous functions are continuous with respect to the induced topologies.

Completeness

Generalising the notion of complete metric space, one can also define completeness for uniform spaces. Instead of working with Cauchy sequences, one works with Cauchy filters (or Cauchy nets).

A Cauchy filter F on a uniform space X is a filter F such that for every entourage U, there exists $A \in F$ with $A \times A \subseteq U$. In other words, a filter is Cauchy if it contains "arbitrarily small" sets. It follows from the definitions that each filter that converges (with respect to the topology defined by the uniform structure) is a Cauchy filter.

A Cauchy filter is called *minimal* if it contains no smaller (i.e., coarser) Cauchy filter (other than itself). It can be shown that every Cauchy filter contains a unique *minimal Cauchy filter*. The neighbourhood filter of each point (the filter consisting of all neighbourhoods of the point) is a minimal Cauchy filter.

Conversely, a uniform space is called complete if every Cauchy filter converges. Any compact Hausdorff space is a complete uniform space with respect to the unique uniformity compatible with the topology.

Complete uniform space enjoy the following important property: if $f: A \to Y$ is a *uniformly continuous* function from a *dense* subset A of a uniform space X into a *complete* uniform space Y, then f can be extended (uniquely) into a uniformly continuous function on all of X. A topological space which can be made into a complete uniform space, whose uniformity induces the original topology, is called a completely uniformizable space.

Hausdorff Completion of a Uniform Space

As with metric spaces, every uniform space X has a Hausdorff completion: that is, there exists a complete Hausdorff uniform space Y and a uniformly continuous map $i: X \to Y$ with the following property:

for any uniformly continuous mapping f of X into a complete Hausdorff uniform space Z, there is a unique uniformly continuous map $g: Y \to Z$ such that $f = gi$.

The Hausdorff completion Y is unique up to isomorphism. As a set, Y can be taken to consist of the *minimal* Cauchy filters on X. As the neighbourhood filter $B(x)$ of each point x in X is a minimal Cauchy filter, the map i can be defined by mapping x to $B(x)$.

The map i thus defined is in general not injective; in fact, the graph of the equivalence relation $i(x) = i(x')$ is the intersection of all entourages of X, and thus i is injective precisely when X is Hausdorff.

The uniform structure on Y is defined as follows: for each *symmetric* entourage V (i.e., such that (x,y) is in V precisely when (y,x) is in V), let $C(V)$ be the set of all pairs (F,G) of minimal Cauchy filters *which have in common at least one V-small set.* The sets $C(V)$ can be shown to form a fundamental system of entourages; Y is equipped with the uniform structure thus defined.

The set $i(X)$ is then a dense subset of Y. If X is Hausdorff, then i is an isomorphism onto $i(X)$, and thus X can be identified with a dense subset of its completion. Moreover, $i(X)$ is always Hausdorff; it is called the Hausdorff uniform space associated with X. If R denotes the equivalence relation $i(x) = i(x')$, then the quotient space X/R is homeomorphic to $i(X)$.

Chapter 4

Distinction Between Geometry and Topology

Pithily, geometry has *local* structure (or infinitesimal), while topology only has *global* structure. Alternatively, geometry has *continuous* moduli, while topology has *discrete* moduli. By examples, an example of geometry is Riemannian geometry, while an example of topology is homotopy theory. The study of metric spaces is geometry, the study of topological spaces is topology. The terms are not used completely consistently: symplectic manifolds are a boundary case, and coarse geometry is global, not local.

Local Versus Global Structure

Differentiable manifolds (of a given dimension) are all locally diffeomorphic (by definition), so there are no local invariants to a differentiable structure (beyond dimension). So differentiable structures on a manifold is an example of topology. By contrast, the curvature of a Riemannian manifold is a local (indeed, infinitesimal) invariant (and is the only local invariant under isometry).

Moduli

If a structure has a discrete moduli (if it has no deformations, or if a deformation of a structure is isomorphic to the original structure), the structure is said to be rigid, and its study (if it is a geometric or topological structure) is topology. If have non-trivial deformations, the structure is said to be flexible, and its study is geometry. The space of homotopy classes of maps is discrete, so studying maps up to homotopy is topology. Similarly, differentiable structures on a manifold is usually a discrete space, and hence an example of topology, but exotic R^4s have continuous moduli of differentiable structures.

Algebraic varieties have continuous moduli spaces, hence their study is algebraic geometry. Note that these are finite-dimensional moduli spaces. The space of Riemannian metrics on a given differentiable manifold is an infinite-dimensional space.

Symplectic Manifolds

Symplectic manifolds are a boundary case, and parts of their study are called symplectic topology and symplectic geometry. By Darboux's theorem, a symplectic manifold has no local structure, which suggests that their study be called topology. By contrast, the space of symplectic structures on a manifold form a continuous moduli, which suggests that their study be called geometry.However, up to isotopy, the space of symplectic structures is discrete (any family of symplectic structures are isotopic).

Topological Degree Theory

In mathematics, topological degree theory is a generalization of the winding number of a curve in the complex plane. It can be used to estimate the number of solutions of an equation, and is closely connected to fixed-point theory. When one solution of an equation is easily found, degree theory can often be used to prove existence of a second, nontrivial, solution. There are different types of degree for different types of maps: e.g. for maps between Banach spaces there is the Brouwer degree in R^n, the Leray-Schauder degree for compact mappings in normed spaces, the coincidence degree and various other types. There is also a degree for continuous maps between manifolds. Topological degree theory has applications in complementarity problems, differential equations, differential inclusions and dynamical systems.

Topological Module

In mathematics, a topological module is a module over a topological ring such that scalar multiplication and addition are continuous.

Examples

A topological vector space is a topological module over a topological field. An abelian topological group can be considered as a topological module over Z, where Z is the ring of integers with the discrete topology. A topological ring is a topological module over each of its subrings.

A more complicated example is the I-adic topology on a ring and its modules. Let I be an ideal of a ring R. The sets of the form $x + I^n$, for all x in R and all positive integers n, form a base for a topology on R that makes R into a topological ring. Then for any left R-module M, the sets of the form $x + I^nM$, for all x in M and all positive integers n, form a base for a topology on M that makes M into a topological module over the topological ring R.

Topological Quantum Field Theory

A topological quantum field theory (or topological field theory or TQFT) is a quantum field theory which computes topological invariants.

Although TQFTs were invented by physicists, they are also of mathematical interest, being related to, among other things, knot theory and the theory of four-manifolds in algebraic topology, and to the theory of moduli spaces in algebraic geometry. Donaldson, Jones, Witten, and Kontsevich have all won Fields Medals for work related to topological field theory.

In condensed matter physics, topological quantum field theories are the low energy effective theories of topologically ordered states, such as fractional quantum Hall states, string-net condensed states, and other strongly correlated quantum liquid states.

Overview

In a topological field theory, the correlation functions do not depend on the metric of spacetime. This means that the theory is not sensitive to changes in the shape of spacetime; if the spacetime warps or contracts, the correlation functions do not change. Consequently, they are topological invariants.

Topological field theories are not very interesting on the flat Minkowski spacetime used in particle physics. Minkowski space can be contracted to a point, so a TQFT on Minkowski space computes only trivial topological invariants.

Consequently, TQFTs are usually studied on curved spacetimes, such as, for example, Riemann surfaces. Most of the known topological field theories are defined on spacetimes of dimension less than five. It seems that a few higher dimensional theories exist, but they are not very well understood.

Quantum gravity is believed to be background-independent (in some suitable sense), and TQFTs provide examples of background independent quantum field theories. This has prompted ongoing theoretical investigation of this class of models.

(Caveat: It is often said that TQFTs have only finitely many degrees of freedom. This is not a fundamental property. It happens to be true in most of the examples that physicists and mathematicians study, but it is not necessary. A topological sigma model with target infinite-dimensional projective space, if such a thing could be defined, would have countably infinitely many degrees of freedom.)

Specific Models

The known topological field theories fall into two general classes: Schwarz-type TQFTs and Witten-type TQFTs. Witten TQFTs are also sometimes referred to as cohomological field theories.

Schwarz-type TQFTs

In Schwarz-type TQFTs, the correlation functions computed by the path integral are topological invariants because the path integral measure and the quantum field observables are explicitly independent of the metric. For instance, in the BF model, the spacetime is a two-dimensional manifold M, the observables are constructed from a two-form F, an auxiliary scalar B, and their derivatives. The action (which determines the path integral) is

$$S = \int_M BF$$

The spacetime metric does not appear anywhere in this theory, so the theory is explicitly topologically invariant. Another, more famous example is Chern-Simons theory, which can be used to compute knot invariants.

Witten-type TQFTs

In Witten-type topological field theories, the topological invariance is more subtle. For example the Lagrangian for the WZW model does depend explicitly on the metric, but one shows by calculation that the expectation value of the partition function and a special class of correlation functions are in fact diffeomorphism invariant.

Mathematical Formulations

Atiyah-Segal Axioms: Atiyah suggested a set of axioms for topological quantum field theory which was inspired by Segal's proposed axioms for conformal field theory, (Atiyah 1988). These axioms have been relatively useful for mathematical treatments of Schwarz-type QFTs, although it isn't clear that they capture the whole structure of Witten-type QFTs. The basic idea is that a TQFT is a functor from a certain category of cobordisms to the category of vector spaces.

There are in fact two different sets of axioms which could reasonably be called the Atiyah axioms. These axioms differ basically in whether or not they study a TQFT defined on a single fixed n-dimensional Riemannian / Lorentzian spacetime M or a TQFT defined on all n-dimensional spacetimes at once.

[ed. *What follows is still in rough draft form and should be regarded suspiciously.*]

The Case of a Fixed Spacetime

Let $Bord_M$ be the category whose morphisms are n-dimensional submanifolds of M and whose objects are connected components of the boundaries of such submanifolds. Regard two morphisms as equivalent if they are homotopic via submanifolds of M, and so form the quotient category $hBord_M$: The objects in $hBord_M$ are the objects of $Bord_M$, and the morphisms of $Bord_M$ are homotopy equivalence classes of morphisms in . A TQFT on M is a symmetric monoidal functor from $Bord_M$ to the category of vector spaces.

Note that cobordisms can, if their boundaries match up, be sewn together to form a new bordism. This is the composition law for morphisms in the cobordism category. Since functors are required to preserve composition, this says that the linear map corresponding to a sewn together morphism is just the composition of the linear map for each piece.

There is an equivalence of categories between the category of 2-dimensional topological quantum field theories and the category of commutative Frobenius algebras.

Topological Rigidity

In mathematics, particularly in topology, a manifold is called topologically rigid if every homotopy-equivalence with any other manifold is homotopic to a homeomorphism.

Motivation

A central problem in topology is determining when two spaces are the same i.e. homeomorphic or diffeomorphic. Constructing a morphism explicitly almost always impractical. If we put further condition on one or both spaces (manifolds) we can exploit this additional structure in order to show that the desired morphism must exist.

Rigidity theorem is about when a fairly weak equivalence between two manifolds (usually a homotopy equivalence) implies the existence of stronger equivalence homeomorphism, diffeomorphism or isometry.

Definition

A closed topological manifold M is called topological rigid if any homotopy equivalence $f : N \to M$ with some manifold N as source and M as target is homotopic to a homeomorphism.

Examples

If closed 2-manifolds M and N are homotopically equivalent then they are homeomorphic. Moreover any homotopy equivalence of closed surfaces deforms to a homeomorphism.

Example

If a closed manifold M^n $(n \neq 3)$ is homotopy-equivalent to S^n then M^n is homeomorphic to S^n.

Rigidity Theorem in Geometry

Definition: A diffeomorphism of flat-Riemannian manifolds is said to be affine iff it carries geodesics to geodesic.

Theorem (Bieberbach)

If $f: M \to N$ is a homotopy equivalence between closed connected Riemannian manifolds then f is homotopic to an affine homeomorphism.

Mostow's Rigidity Theorem

Theorem: Let M and N be compact, locally symmetric Riemannian manifolds with everywhere non-positive curvature having no closed one or two dimensional geodesic subspace which are direct factor locally. If $f: M \to N$ is a homotopy equivalence then f is homotopic to an isometry.

Theorem (Mostow's theorem for hyperbolic n-manifolds, $n \geq 3$): If M and N are complete hyperbolic n-manifolds, $n \geq 3$ with finite volume and $f: M \to N$ is a homotopy equivalence then f is homotopic to an isometry. These results are named after George Mostow.

Algebraic Form

Let Γ and Δ be discrete subgroups of the isometry group of hyperbolic n-space H, where $n \geq 3$, whose quotients H/Γ and H/Δ have finite volume. If Γ and Δ are isomorphic as discrete groups then they are conjugate.

Remarks

(1) In the 2-dimensional case any manifold of genus at least two has a hyperbolic structure. Mostow's rigidity theorem does not apply in this case. In fact, there are many hyperbolic structures on any such manifold; each such structure corresponds to a point in Teichmuller space.

(2) On the other hand if M and N are 2-manifolds of finite volume then it is easy to show that they are homeomorphic exactly when their fundamental groups are the same.

Application

The group of isometries of a finite-volume hyperbolic n-manifold M (for $n \geq 3$) is finite and isomorphic to $\pi_1(M)$.

Topology Optimization

Topology optimisation is a mathematical approach that optimises material layout within a given design space, for a given set of loads and boundary conditions such that the resulting layout meets a prescribed set of performance targets. Using topology optimisation, engineers can find the best concept design that meets the design requirements. Topology optimisation has been implemented through the use of finite element methods for the analysis, and optimisation techniques based on the method of moving asymptotes, genetic algorithms, optimality criteria method, level sets, and topological derivatives. Topology optimisation is used at the concept level of the design process to arrive at a conceptual design proposal that is then fine tuned for performance and manufacturability. This replaces time consuming and costly design iterations and hence reduces design development time and overall cost while improving design performance.

In some cases, proposals from a topology optimisation, although optimal, may be expensive or infeasible to manufacture. These challenges can be overcome through the use of manufacturing constraints in the topology optimisation problem formulation. Using manufacturing constraints, the optimisation yields engineering designs that would satisfy practical manufacturing requirements. In some cases Additive manufacturing technologies are used to manufacture complex optimized shapes that would otherwise need manufacturing constraints.

Problem Statement

Mathematically one can pose a generic problem as follows:

$$\min_{\rho} \int_{\Omega} \phi(\rho) \mathrm{d}\Omega$$

subject to:

$$\rho \in \{0,1\}$$

Design Constraints

Governing Differential Equation

The problem statement includes the following

Objective Functional $\left(\int_{\Omega} \phi(\rho) \mathrm{d}\Omega\right)$***:*** This is the goal of the optimization study which is to be minimised over the selection field. For example, one would want to minimise the compliance of a structure to increase structural stiffness.

Design Space (Ω)**:** Design space is the allowable volume within which the design can exist. Assembly and packaging requirements, human and tool accessibility are some of the factors that need to be considered in identifying this space.

With the definition of the design space, regions or components in the model that cannot be modified during the course of the optimisation are considered as non-design regions.

The Discrete Selection Field (ρ)**:** This is the field over which the discrete optimisation is to be performed. It selects or deselects a point on the design space to further the design objective. By selection it has to take the value 1and by de-selection it has to take the tvalue 0.

Design constraints: These are design criteria that need to satisfied. These could include material availability constraints, displacement constraints, etc.

Governing Differential Equation: This is the one that governs the physics of the structure to be built. For example the elasticity equation in the case of stiff structures would be the governing differential equation.

Implementation Methodology

The optimisation problem posed above implies the need for a discrete optimisation over every point in domain Ω. But, implementation of such a scheme is not yet practicable. So, people adopt continuous optimisation strategies, where they assume that the variable ρ varies continuously over the domain $(0,1)$. This problem is then solved over a set of finite elements obtained from meshing of the domain.

Topology Optimization for Stiff Structures

A stiff structure is one that has the least possible displacement when given certain set of boundary conditions. A global measure of the displacements is the strain energy (also called compliance) of the structure under the prescribed boundary conditions.

The lower the strain energy the higher the stiffness of the structure. So, the problem statement involves the objective functional of the strain energy which has to be minimized.

Now the objective functional should be chosen as a function of the selection field ρ. So in literature, people have interpolated the material properties in terms of the selection field. A widely used interpolation scheme is called the Solid Isotropic Material with Penalization (SIMP). This interpolation is essentially a power law $E = E_0 + \rho^n E_1$ that interpolates the Young's modulus of the material to the scalar selection field.

The value of varies between $[0,3]$ in general. This has been shown to confirm to micro-structure of the materials. So one could view topology optimization to be a process of selection of micro-structure at every point in space so that an objective functional is minimized.

On a broad level, one can visualize that more the material, lesser will be the deflection as there is more material to resist the loads. So, the optimization requires an opposing constraint, the constraint of the volume. This is in reality a cost factor too and we would not want to spend a lot of money on the material. To obtain the total material utilized an integration of the selection field over the volume can be done.

Finally the elasticity governing differential equations are plugged in so as to get the final problem statement.

$$\min_{\rho} \int_{\Omega} \frac{1}{2} \boldsymbol{\sigma} : \boldsymbol{\varepsilon} d\Omega$$

subject to:

$$\rho \in [0,1]$$

$$\int_{\Omega} \rho d\Omega \leq V^*$$

$$\nabla \cdot \boldsymbol{\sigma} + \mathbf{F} = \mathbf{0}$$

$$\boldsymbol{\sigma} = \mathbf{C} : \boldsymbol{\varepsilon}$$

But, a straight forward implementation in the Finite Element Framework of such a problem is still infeasible owing to issues such as:

1. Mesh dependency — Mesh Dependency means that the design obtained on one mesh is not the one that will be obtained on an another mesh. The features of the design become more intricate as the mesh gets refined.

2. Numerical instabilities — The selection of region in the form of a chess board.

Figure: Checker Board Patterns are shown in this result.

Some techniques called as Filtering based on Image Processing are currently being used to alleviate some of these issues.

Figure: Topology Optimization result when filtering is used

Distinction

Topology optimization is distinct from shape optimization since typically shape optimisation methods work in a subset of allowable shapes which have fixed topological properties, such as having a fixed number of holes in them. Therefore topology optimisation is used to generate concepts and shape optimisation is used to fine-tune a chosen design topology. There are various methods used to perform topology optimisation:

- Solid Isotropic Material with Penalisation (SIMP),
- Evolutionary Structural Optimisation (ESO),
- Topological derivatives.

Triangulation (Topology)

In mathematics, topology generalizes the notion of triangulation in a natural way as follows:

A triangulation of a topological space X is a simplicial complex K, homeomorphic to X, together with a homeomorphism $h{:}K \rightarrow X$.

Triangulation is useful in determining the properties of a topological space. For example, one can compute homology and cohomology groups of a triangulated space using simplicial homology and cohomology theories instead of more complicated homology and cohomology theories.

Piecewise Linear Manifold

In mathematics, a piecewise linear (PL) manifold is a topological manifold together with a piecewise linear structure on it. Such a structure can be defined by means of an atlas, such that one can pass from chart to chart in it by piecewise linear functions. This is slightly stronger than the topological notion of a triangulation. An isomorphism of PL manifolds is called a PL homeomorphism.

Relation to other Categories of Manifolds

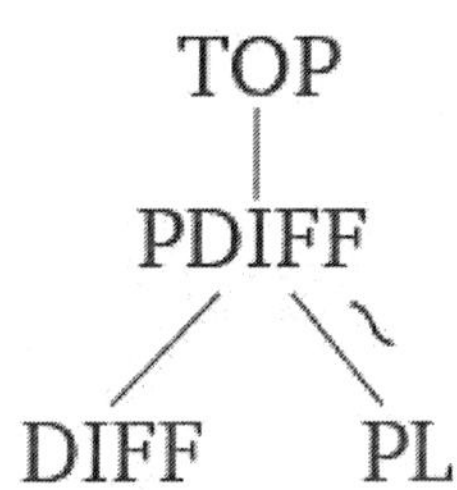

Figure: *PDIFF serves to relate DIFF and PL, and it is equivalent to PL.*

PL, or more precisely PDIFF, sits between DIFF (the category of smooth manifolds) and TOP (the category of topological manifolds): it is categorically "better behaved" than DIFF – for example, the Generalized Poincaré conjecture is true in PL (with the possible exception of dimension 4, where it is equivalent to DIFF), but is false generally in DIFF – but is "worse behaved" than TOP, as elaborated in surgery theory.

Smooth Manifolds

Smooth manifolds have canonical PL structures – they are uniquely *triangulizable,* by Whitehead's theorem on triangulation (Whitehead 1940) – but PL manifolds do not always have smooth structures – they are not always *smoothable.* This relation can be elaborated by introducing the category PDIFF, which contains both DIFF and PL, and is equivalent to PL.

One way in which PL is better behaved than DIFF is that one can take cones in PL, but not in DIFF – the cone point is acceptable in PL. A consequence is that the Generalized Poincaré conjecture is true in PL for dimensions greater than four – the proof is to take a homotopy sphere, remove two balls, apply the h-cobordism theorem to conclude that this is a cylinder, and then attach cones to recover a sphere. This last step works in PL but not in DIFF, giving rise to exotic spheres.

Topological Manifolds

Not every topological manifold admits a PL structure, and of those that do, the PL structure need not be unique – it can have infinitely many. This is elaborated at Hauptvermutung.

The obstruction to placing a PL structure on a topological manifold is the Kirby–Siebenmann class. To be precise, the Kirby-Siebenmann class is the obstruction to placing a PL-structure on M x R and in dimensions n > 4 this ensures that M has a PL-structure.

Real Algebraic Sets

An A-structure on a PL manifold is a structure which gives an inductive way of resolving the PL manifold to a smooth manifold. Compact PL manifolds admit A-structures (Akbulut & Taylor). Compact PL manifolds are homeomorphic to real-algebraic sets (Akbulut & King). Put another way, A-category sits over the PL-category as a richer category with no obstruction to lifting, that is BA —> BPL is a product fibration with BA = BPL x PL/A, and PL manifolds are real algebraic sets because A-manifolds are real algebraic sets.

Combinatorial manifolds and digital manifolds

- A combinatorial manifold is a kind of manifold which is discretization of a manifold. It usually means a piecewise linear manifold made by simplicial complexes.
- A digital manifold is a special kind of combinatorial manifold which is defined in digital space.

Explicit Methods of Triangulation

An important special case of topological triangulation is that of two-dimensional surfaces, or closed 2-manifolds. There is a standard proof that smooth closed surfaces can be triangulated. Indeed, if the surface is given a Riemannian metric, each point x is contained inside a small convex geodesic triangle lying inside a normal ball with centre x. The interiors of finitely many of the triangles will cover the surface; since edges of different triangles either coincide or intersect transversally, this finite set of triangles can be used iteratively to construct a triangulation.

Another simple procedure for triangulating differentiable manifolds was given by Hassler Whitney in 1957, based on his embedding theorem. In fact, if X is a closed n-submanifold of R^m, subdivide a cubical lattice in R^m into simplices to give a triangulation of R^m. By taking the mesh of the lattice small enough and slightly moving finitely many of the vertices, the triangulation will be in *general position* with respect to X: thus no simplices of dimension $< s=m-n$ intersect X and each s-simplex intersecting X

- does so in exactly one interior point;
- makes a strictly positive angle with the tangent plane;
- lies wholly inside some tubular neighbourhood of X.

These points of intersection and their barycentres (corresponding to higher dimensional simplices intersecting X) generate an n-dimensional simplicial subcomplex in R^m, lying wholly inside the tubular neighbourhood. The triangulation is given by the projection of this simplicial complex onto X.

Graphs on Surfaces

A *Whitney triangulation* or *clean triangulation* of a surface is an embedding of a graph onto the surface in such a way that the faces of the embedding are exactly the cliques of the graph (Hartsfeld and Gerhard Ringel 1981; Larrión et al. 2002; Malniè and Mohar 1992). Equivalently, every face is a triangle, every triangle is a face, and the graph is not itself a clique. The clique complex of the graph is then homeomorphic to the surface. The 1-skeletons of Whitney triangulations are exactly the locally cyclic graphs other than K_4.

Topological Combinatorics

The discipline of combinatorial topology used combinatorial concepts in topology and in the early 20th century this gradually turned into the field of algebraic topology.

In 1978 the situation was reversed when methods from algebraic topology were used to solve a problem in combinatorics when Loszló Lovász proved the Kneser conjecture, thus beginning the new study of topological combinatorics. Lovász's proof used the Borsuk-Ulam theorem and this theorem retains a prominent role in this new field. This theorem has many equivalent versions and analogs and has been used in the study of fair division problems.

In another application of homological methods to graph theory Lovász proved both the undirected and directed versions of a conjecture of Frank: Given a k-connected graph G, k points $v_1,...,v_k \in V(G)$, and k positive integers $n_1,n_2,...,n_k$ that sum up to $|V(G)|$, there exists a partition $\{V_1,...,V_k\}$ of $V(G)$ such that $v_i \in V_i$, $|V_i|=n_i$ and V_i spans a connected subgraph.

The most notable application of topological combinatorics has been to graph colouring problems. Also in 1987 the necklace problem was solved by Noga Alon.

It has also been used to study complexity problems in linear decision tree algorithms and the Aanderaa–Karp–Rosenberg conjecture. Other areas include topology of partially ordered sets and bruhat orders. Also methods from differential topology now have a combinatorial analog in discrete Morse theory.

Geospatial Topology

Geospatial topology studies the rules concerning the relationships between the points, lines, and polygons that represent the features of a geographic region. For example, where two polygons represent adjacent counties, typical topological rules would require that the counties share a common boundary with no gaps and no overlaps. Similarly, it would be nonsense to allow two polygons representing lakes to overlap.

The ESRI White Paper *GIS Topology* explains that topology operations are used to manage shared geometry, define and enforce data integrity rules, support topological relationship queries and navigation, and build more complex shapes such as polygons, from primitive ones such as lines. A *GIS for Educators* worksheet at Linfiniti adds the detection and correction of digitising errors and carrying out network analysis. Topological error correction is explained in more detail in a paper by Ubeda and Egenhofer.

Unlike GML, topologies are not directly represented in ESRI shapefiles which store individual geometric objects in isolation. Topological processing can, however, be undertaken in GIS software such as QGIS or could in principle be enforced using integrity constraints in a GIS-enabled DBMS such as PostGIS.

However, as Riedemann (2004) explains, topological operators are inherently complex and their implementation requires care to be taken with usability and conformance to standards. Oracle and PostGIS provide fundamental toplogical operators allowing applications to test for *"such relationships as contains, inside, covers, covered by, touch, and overlap with boundaries intersecting."*

Unlike the PostGIS documentation, the Oracle documentation draws a distinction between *"topological relationships [which] remain constant when the coordinate space is deformed, such as by twisting or stretching"* and *"relationships that are not topological [which] include length of, distance between, and area of."*

These operators are leveraged by applications to ensure that data is stored an processed in a topologically correct fashion.

Lamination (Topology)

In topology, a branch of mathematics, a lamination is a :

- "A topological space partitioned into subsets"
- decoration (a structure or property at a point) of a manifold in which some subset of the manifold is partitioned into sheets of some lower dimension, and the sheets are locally parallel.

Lamination of surface is a partition of closed subset of surface into unions of smooth curves. It may or may not be possible to fill the gaps in a lamination to make a foliation.

Examples

- A geodesic lamination of a 2-dimensional hyperbolic manifold is a closed subset together with a foliation of this closed subset by geodesics. These are used in Thurston's classification of elements of the mapping class group and in his theory of earthquake maps.
- Quadratic laminations, which remain invariant under the angle doubling map. These laminations are associated with quadratic maps . It is a closed collection of chords in the unit disc . It is also topological model of Mandelbrot or Julia set.

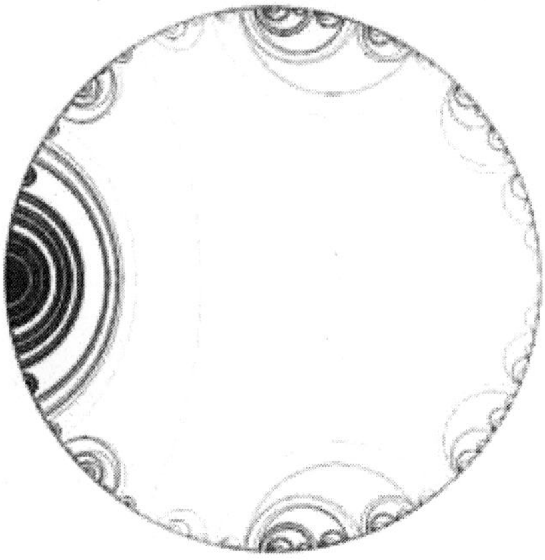

***Figure:** Lamination associated with Mandelbrot set*

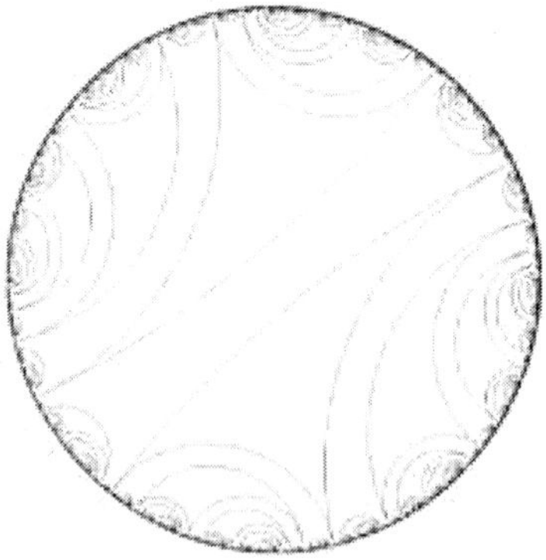

***Figure:** Lamination of rabbit Julia set.*

Loop (Topology)

A loop in mathematics, in a topological space X is a path f from the unit interval $I = [0,1]$ to X such that $f(0) = f(1)$. In other words, it is a path whose initial point is equal to the terminal point.

A loop may also be seen as a continuous map f from the unit circle S^1 into X, because S^1 may be regarded as a quotient of I under the identification $0 \sim 1$.

The set of all loops in X forms a space called the loop space of X.

Strong Operator Topology

In functional analysis, a branch of mathematics, the strong operator topology, often abbreviated SOT, is the weakest locally convex topology on the set of bounded operators on a Hilbert space (or, more generally, on a Banach space) such that the evaluation map sending an operator T to the real number $\|Tx\|$ is continuous for each vector x in the Hilbert space.

The SOT is stronger than the weak operator topology and weaker than the norm topology. The SOT lacks some of the nicer properties that the weak operator topology has, but being stronger, things are sometimes easier to prove in this topology. It is more natural too, since it is simply the topology of pointwise convergence for an operator.

The SOT topology also provides the framework for the measurable functional calculus, just as the norm topology does for the continuous functional calculus.

The linear functionals on the set of bounded operators on a Hilbert space that are continuous in the SOT are precisely those continuous in the WOT. Because of this, the closure of a convex set of operators in the WOT is the same as the closure of that set in the SOT.

It should also be noted that the above language translates into convergence properties of Hilbert space operators.

One especially observes that for a complex Hilbert space, by way of the polarization identity, one easily verifies that Strong Operator convergence implies Weak Operator convergence.

Strong Topology (Polar Topology)

In functional analysis and related areas of mathematics the strong topology is the finest polar topology, the topology with the most open sets, on a dual pair. The coarsest polar topology is called weak topology.

Definition

Given a dual pair $(X,Y,\langle,\rangle)$ the strong topology $\beta(Y,X)$ on Yis the polar topology defined by using the family of all sets in X where the polar set in is absorbent.

Examples

- Given a normed vector space Xand its continuous dual Xthen $\beta(X',X)$-topology on Xis identical to the topology induced by the operator norm. Conversely $\beta(X,X')$-topology on Xis identical to the topology induced by the norm.

Properties

In barrelled spaces the strong topology is identical to the Mackey topology.

Trivial Topology

In topology, a topological space with the trivial topology is one where the only open sets are the empty set and the entire space. Such a space is sometimes called an indiscrete space, and its topology sometimes called an indiscrete topology. Intuitively, this has the consequence that all points of the space are "lumped together" and cannot be distinguished by topological means; it belongs to a pseudometric space in which the distance between any two points is zero.

The trivial topology is the topology with the least possible number of open sets, since the definition of a topology requires these two sets to be open. Despite its simplicity, a space *X* with more than one element and the trivial topology lacks a key desirable property: it is not a T_0 space.

Other properties of an indiscrete space *X*—many of which are quite unusual—include:

- The only closed sets are the empty set and *X*.

- The only possible basis of X is $\{X\}$.
- If X has more than one point, then since it is not T_0, it does not satisfy any of the higher T axioms either. In particular, it is not a Hausdorff space. Not being Hausdorff, X is not an order topology, nor is it metrizable.
- X is, however, regular, completely regular, normal, and completely normal; all in a rather vacuous way though, since the only closed sets are $\varnothing$ and X.
- X is compact and therefore paracompact, Lindelöf, and locally compact.
- Every function whose domain is a topological space and codomain X is continuous.
- X is path-connected and so connected.
- X is second-countable, and therefore is first-countable, separable and Lindelöf.
- All subspaces of X have the trivial topology.
- All quotient spaces of X have the trivial topology
- Arbitrary products of trivial topological spaces, with either the product topology or box topology, have the trivial topology.
- All sequences in X converge to every point of X. In particular, every sequence has a convergent subsequence (the whole sequence), thus X is sequentially compact.
- The interior of every set except X is empty.
- The closure of every non-empty subset of X is X. Put another way: every non-empty subset of X is dense, a property that characterizes trivial topological spaces.
- If S is any subset of X with more than one element, then all elements of X are limit points of S. If S is a singleton, then every point of $X \setminus S$ is still a limit point of S.
- X is a Baire space.
- Two topological spaces carrying the trivial topology are homeomorphic iff they have the same cardinality.

In some sense the opposite of the trivial topology is the discrete topology, in which every subset is open.

The trivial topology belongs to a uniform space in which the whole cartesian product $X \times X$ is the only entourage.

Let Top be the category of topological spaces with continuous maps and Set be the category of sets with functions. If F : Top $\to$ Set is the functor that assigns to each topological space its underlying set (the so-called forgetful functor), and G : Set $\to$ Top is the functor that puts the trivial topology on a given set, then G is right adjoint to F. (The functor H : Set $\to$ Top that puts the discrete topology on a given set is *left adjoint* to F.)

Differential Topology

In mathematics, differential topology is the field dealing with differentiable functions on differentiable manifolds. It is closely related to differential geometry and together they make up the geometric theory of differentiable manifolds.

Description

Differential topology considers the properties and structures that require only a smooth structure on a manifold to be defined. Smooth manifolds are 'softer' than manifolds with extra geometric structures, which can act as obstructions to certain types of equivalences and deformations that exist in differential topology. For instance, volume and Riemannian curvature are invariants that can distinguish different geometric structures on the same smooth manifold—that is, one can smoothly "flatten out" certain manifolds, but it might require distorting the space and affecting the curvature or volume.

On the other hand, smooth manifolds are more rigid than the topological manifolds. John Milnor discovered that some spheres have more than one smooth structure—see exotic sphere and Donaldson's theorem. Kervaire exhibited topological manifolds with no smooth structure at all. Some constructions of smooth manifold theory, such as the existence of tangent bundles, can be done in the topological setting with much more work, and others cannot.

One of the main topics in differential topology is the study of special kinds of smooth mappings between manifolds, namely immersions and submersions, and the intersections of submanifolds via transversality. More generally one is interested in properties and invariants of smooth manifolds which are carried over by diffeomorphisms, another special kind of smooth mapping. Morse theory is another branch of differential topology, in which topological information about a manifold is deduced from changes in the rank of the Jacobian of a function.

Differential Topology Versus Differential Geometry

Differential topology and differential geometry are first characterized by their *similarity*. They both study primarily the properties of differentiable manifolds, sometimes with a variety of structures imposed on them.

One major difference lies in the nature of the problems that each subject tries to address. In one view, differential topology distinguishes itself from differential geometry by studying primarily those problems which are *inherently global*. Consider the example of a coffee cup and a donut. From the point of view of differential topology, the donut and the coffee cup are *the same* (in a sense). This is an inherently global view, though, because there is no way for the differential topologist to tell whether the two objects are the same (in this sense) by looking at just a tiny (*local*) piece of either of them. He must have access to each entire (*global*) object.

From the point of view of differential geometry, the coffee cup and the donut are *different* because it is impossible to rotate the coffee cup in such a way that its configuration matches that of the donut. This is also a global way of thinking about the problem. But an important distinction is that the geometer does not need the entire object to decide this. By looking, for instance, at just a tiny piece of the handle, he can decide that the coffee cup is different from the donut because the handle is thinner (or more curved) than any piece of the donut.

To put it succinctly, differential topology studies structures on manifolds which, in a sense, have no interesting local structure. Differential geometry studies structures on manifolds which do have an interesting local (or sometimes even infinitesimal) structure.

More mathematically, for example, the problem of constructing a diffeomorphism between two manifolds of the same dimension is inherently global since *locally* two such manifolds are always diffeomorphic. Likewise, the problem of computing a quantity on a manifold which is invariant under differentiable mappings is inherently global, since any local invariant will be *trivial* in the sense that it is already exhibited in the topology of R^n. Moreover, differential topology does not restrict itself necessarily to the study of diffeomorphism. For example, symplectic topology — a subbranch of differential topology — studies global properties of symplectic manifolds. Differential geometry concerns itself with problems — which may be local *or* global — that always have some non-trivial local properties. Thus differential geometry may study differentiable manifolds equipped with a *connection*, a *metric* (which may be Riemannian, pseudo-Riemannian, or Finsler), a special sort of *distribution* (such as a CR structure), and so on.

This distinction between differential geometry and differential topology is blurred, however, in questions specifically pertaining to local diffeomorphism invariants such as the tangent space at a point. Differential topology also deals with questions like these, which specifically pertain to the properties of differentiable mappings on R^n (for example the tangent bundle, jet bundles, the Whitney extension theorem, and so forth).

Nevertheless, the distinction becomes clearer in abstract terms. Differential topology is the study of the (infinitesimal, local, and global) properties of structures on manifolds having no non-trivial local moduli, whereas differential geometry is the study of the (infinitesimal, local, and global) properties of structures on manifolds having non-trivial local moduli.

Chapter 5

Metrization Theorem

In topology and related areas of mathematics, a metrizable space is a topological space that is homeomorphic to a metric space. That is, a topological space (X, τ) is said to be metrizable if there is a metric

$$d : X \times X \to [0, \infty)$$

such that the topology induced by *d* is t. Metrization theorems are theorems that give sufficient conditions for a topological space to be metrizable.

Properties

Metrizable spaces inherit all topological properties from metric spaces. For example, they are Hausdorff paracompact spaces (and hence normal and Tychonoff) and first-countable. However, some properties of the metric, such as completeness, cannot be said to be inherited. This is also true of other structures linked to the metric. A metrizable uniform space, for example, may have a different set of contraction maps than a metric space to which it is homeomorphic.

Metrization Theorems

The first really useful metrization theorem was Urysohn's metrization theorem. This states that every Hausdorff second-countable regular space is metrizable. So, for example, every second-countable manifold is metrizable. (Historical note: The form of the theorem shown here was in fact proved by Tychonoff in 1926. What Urysohn had shown, in a paper published posthumously in 1925, was that every second-countable *normal* Hausdorff space is metrizable.)

Several other metrization theorems follow as simple corollaries to Urysohn's Theorem. For example, a compact Hausdorff space is metrizable if and only if it is second-countable.

Urysohn's Theorem can be restated as: A topological space is separable and metrizable if and only if it is regular, Hausdorff and second-countable. The Nagata-Smirnov metrization theorem extends this to the non-separable case. It states that a topological space is metrizable if and only if it is regular, Hausdorff and has a σ-locally finite base.

A σ-locally finite base is a base which is a union of countably many locally finite collections of open sets.

Separable metrizable spaces can also be characterised as those spaces which are homeomorphic to a subspace of the Hilbert cube $[0,1]^{\mathbb{N}}$,, i.e. the countably infinite product of the unit interval (with its natural subspace topology from the reals) with itself, endowed with the product topology.

A space is said to be locally metrizable if every point has a metrizable neighbourhood. Smirnov proved that a locally metrizable space is metrizable if and only if it is Hausdorff and paracompact. In particular, a manifold is metrizable if and only if it is paracompact.

Examples of Non-Metrizable Spaces

Non-normal spaces cannot be metrizable; important examples include

- the Zariski topology on an algebraic variety or on the spectrum of a ring, used in algebraic geometry,
- the topological vector space of all functions from the real line R to itself, with the topology of pointwise convergence.
- the Strong operator topology on the set of unitary operators on a Hilbert Space *H* (often denoted U(H))

The real line with the lower limit topology is not metrizable. The usual distance function is not a metric on this space because the topology it determines is the usual topology, not the lower limit topology. This space is Hausdorff, paracompact and first countable. The long line is locally metrizable but not metrizable; in a sense it is "too long".

Polish Space

In the mathematical discipline of general topology, a Polish space is a separable completely metrizable topological space; that is, a space homeomorphic to a complete metric space that has a countable dense subset. Polish spaces are so named because they were first extensively studied by Polish topologists and logicians — Sierpiñski, Kuratowski, Tarski and others. However, Polish spaces are primarily studied today because they are the primary setting for descriptive set theory, including the study of Borel equivalence relations.

Common examples of Polish spaces are the real line, any separable Banach space, the Cantor space, and Baire space. Additionally, some spaces that are not complete metric spaces in the usual metric may be Polish; e.g., the open interval (0, 1) is Polish.

Between any two uncountable Polish spaces, there is a Borel isomorphism; that is, a bijection that preserves the Borel structure. In particular, every uncountable Polish space has the cardinality of the continuum. Lusin spaces, Suslin spaces, and Radon spaces are generalizations of Polish spaces.

Properties

1. (Alexandrov's theorem) If X is Polish then so is any G_δ subset of X.
2. (Cantor–Bendixson theorem) If X is Polish then any closed subset of X can be written as the disjoint union of a perfect subset and a countable subset.

The converse of Alexandrov's theorem is true as well: if a subspace S of a Polish space X is Polish, then it is a G_δ subset of X.

Characterisation

There are numerous characterisations that tell when a second countable topological space is metrizable, such as Urysohn's metrization theorem. The problem of determining whether a metrizable space is completely metrizable is more difficult. Topological spaces such as the open unit interval (0,1) can be given both complete metrics and incomplete metrics generating their topology.

There is a characterisation of complete separable metric spaces in terms of a game known as the strong Choquet game. A separable metric space is completely metrizable if and only if the second player has a winning strategy in this game.

A second characterisation follows from Alexandrov's theorem. It states that a separable metric space is completely metrizable if and only if it is a Gδ subset of its completion in the original metric.

Polish Metric Spaces

Although Polish spaces are metrizable, they are not in and of themselves metric spaces; each Polish space admits many complete metrics giving rise to the same topology, but no one of these is singled out or distinguished. A Polish space with a distinguished complete metric is called a *Polish metric space*. An alternative approach, equivalent to the one given here, is first to define "Polish metric space" to mean "complete separable metric space", and then to define a "Polish space" as the topological space obtained from a Polish metric space by forgetting the metric.

Generalizations of Polish Spaces

Lusin Spaces: A Lusin space is a topological space such that some weaker topology makes it into a Polish space.

There are many ways to form Lusin spaces. In particular:

- Every Polish space is Lusin.
- A subspace of a Lusin space is Lusin if and only if it is a Borel set.
- Any countable union or intersection of Lusin subspaces of a Hausdorff space is Lusin.
- The product of a countable number of Lusin spaces is Lusin.
- The disjoint union of a countable number of Lusin spaces is Lusin.

Suslin Spaces

A Suslin space is the image of a Polish space under a continuous mapping. So every Lusin space is Suslin. In a Polish space, a subset is a Suslin space if and only if it is a Suslin set (an image of the Suslin operation).

Radon Spaces

A Radon space is a topological space such that every finite Borel measure is inner regular (so a Radon measure). Every Suslin space is Radon.

Polish Groups

A Polish group is a topological group G regarded as a topological space which is itself a Polish space. A remarkable fact about Polish groups is that Baire-measurable (i.e., the preimage of any open set has the property of Baire) homomorphisms between them are automatically continuous. (Pettis in B. J. Pettis, 'On continuity and openness of homomorphisms in topological groups', Ann. of Math. vol. 51 (1950) 293–308, MR 38358)

Stone Duality

In mathematics, there is an ample supply of categorical dualities between certain categories of topological spaces and categories of partially ordered sets. Today, these dualities are usually collected under the label Stone duality, since they form a natural generalization of Stone's representation theorem for Boolean algebras. These concepts are named in honour of Marshall Stone. Stone-type dualities also provide the foundation for pointless topology and are exploited in theoretical computer science for the study of formal semantics.

This article gives pointers to special cases of Stone duality and explains a very general instance thereof in detail.

Overview of Stone-Type Dualities

Probably the most general duality which is classically referred to as "Stone duality" is the duality between the category Sob of sober spaces with continuous functions and the category SFrm of spatial frames with appropriate frame homomorphisms. The dual category of SFrm is the category of locales denoted by SLoc.

The categorical equivalence of Sob and SLoc is the basis for the mathematical area of pointless topology, that is devoted to the study of Loc - the category of all locales of which SLoc is a full subcategory. The involved constructions are characteristic for this kind of duality, and are detailed below.

Now one can easily obtain a number of other dualities by restricting to certain special classes of sober spaces:

- The category CohSp of coherent sober spaces (and coherent maps) is equivalent to the category CohLoc of coherent (or spectral) locales (and coherent maps), on the assumption of the Boolean prime ideal theorem (in fact, this statement is equivalent to that assumption). The significance of this result stems from the fact that CohLoc in turn is dual to the category DLat of distributive lattices. Hence, DLat is dual to CohSp - one obtains Stone's representation theorem for distributive lattices.
- When restricting further to coherent sober spaces which are Hausdorff, one obtains the category Stone of so-called Stone spaces. On the side of DLat, the restriction yields the subcategory Bool of Boolean algebras. Thus one obtains Stone's representation theorem for Boolean algebras.
- Stone's representation for distributive lattices can be extended via an equivalence of coherent spaces and Priestley spaces (ordered topological spaces, that are compact and totally order-disconnected). One obtains a representation of distributive lattices via ordered topologies: Priestley's representation theorem for distributive lattices.

Many other Stone-type dualities could be added to these basic dualities.

Duality of Sober Spaces and Spatial Locales

This section motivates and explains one of the most basic constructions of Stone duality: the duality between topological spaces which are *sober* and frames (i.e. complete Heyting algebras) which are *spatial*. This classical piece of mathematics requires a substantial amount of abstraction that usually tends to puzzle beginners.

It should therefore be considered as graduate level mathematics. Some prior exposure to the basics of category theory is recommended, although a deep understanding of the concepts of adjunction and duality may well arise from examples such as the result below. Furthermore, concepts of topology and order theory are naturally involved as well, where the latter is probably more important for a thorough understanding.

The Lattice of Open Sets

The starting point for the theory is the fact that every topological space is characterised by a set of points X and a system $\Omega(X)$ of open sets of elements from X, i.e. a subset of the powerset of X. It is known that $\Omega(X)$ has certain special properties: it is a complete lattice within which suprema and infima of finite sets are given by set unions and finite set intersections, respectively. Furthermore, it contains both X and the empty set. Since the embedding of $\Omega(X)$ into the powerset lattice of X preserves finite infima and arbitrary suprema, $\Omega(X)$ inherits the following distributivity law:

$$x \wedge \bigvee S = \bigvee\{x \wedge s : s \in S\},$$

for every element (open set) x and every subset S of $\Omega(X)$. Hence $\Omega(X)$ is not an arbitrary complete lattice but a *complete Heyting algebra* (also called *frame* or *locale* - the various names are primarily used to distinguish several categories that have the same class of objects but different morphisms: frame morphisms, locale morphisms and homomorphisms of complete Heyting algebras). Now an obvious question is: To what extent is a topological space characterised by its locale of open sets?

As already hinted at above, one can go even further. The category Top of topological spaces has as morphisms the continuous functions, where a function f is continuous if the inverse image $f^{-1}(O)$ of any open set in the codomain of f is open in the domain of f.

Thus any continuous function f from a space X to a space Y defines an inverse mapping f^{-1} from $\Omega(Y)$ to $\Omega(X)$. Furthermore, it is easy to check that f^{-1} (like any inverse image map) preserves finite intersections and arbitrary unions and therefore is a *morphism of frames*. If we define $\Omega(f) = f^{-1}$ then Ω becomes a contravariant functor from the category Top to the category Frm of frames and frame morphisms. Using the tools of category theory, the task of finding a characterisation of topological spaces in terms of their open set lattices is equivalent to finding a functor from Frm to Top which is adjoint to Ω.

Points of a Locale

The goal of this section is to define a functor pt from Frm to Top that in a certain sense "inverts" the operation of Ω by assigning to each locale L a set of points pt(L) (hence the notation pt) with a suitable topology. But how can we recover the set of points just from the locale, though it is not given as a lattice of sets?

It is certain that one cannot expect in general that pt can reproduce all of the original elements of a topological space just from its lattice of open sets - for example all sets with the indiscrete topology yield (up to isomorphism) the same locale, such that the information on the specific set is no longer present.

However, there is still a reasonable technique for obtaining "points" from a locale, which indeed gives an example of a central construction for Stone-type duality theorems.

Let us first look at the points of a topological space X. One is usually tempted to consider a point of X as an element x of the set X, but there is in fact a more useful description for our current investigation.

Any point x gives rise to a continuous function p_x from the one element topological space 1 (all subsets of which are open) to the space X by defining $p_x(1) = x$. Conversely, any function from 1 to X clearly determines one point: the element that it "points" to.

Therefore the set of points of a topological space is equivalently characterised as the set of functions from 1 to X. When using the functor Ω to pass from Top to Frm, all set-theoretic elements of a space are lost, but - using a fundamental idea of category theory - one can as well work on the function spaces.

Indeed, any "point" p_x: $1 \to X$ in Top is mapped to a morphism $\Omega(p_x)$: $\Omega(X) \to \Omega(1)$. The open set lattice of the one-element topological space $\Omega(1)$ is just (isomorphic to) the two-element locale $2 = \{ 0, 1 \}$ with $0 < 1$.

After these observations it appears reasonable to define the set of points of a locale L to be the set of frame morphisms from L to 2. Yet, there is no guarantee that every point of the locale $\Omega(X)$ is in one-to-one correspondence to a point of the topological space X (consider again the indiscrete topology, for which the open set lattice has only one "point").

Before defining the required topology on pt(X), it is worthwhile to clarify the concept of a point of a locale further. The perspective motivated above suggests to consider a point of a locale L as a frame morphism p from L to 2.

But these morphisms are characterised equivalently by the inverse images of the two elements of 2. From the properties of frame morphisms, one can derive that $p^{-1}(0)$ is a lower set (since p is monotone), which contains a greatest element $a_p = \bigvee p^{-1}(0)$ (since p preserves arbitrary suprema).

In addition, the principal ideal $p^{-1}(0)$ is a prime ideal since p preserves finite infima and thus the principal a_p is a meet-prime element. Now the set-inverse of $p^{-1}(0)$ given by $p^{-1}(1)$ is a completely prime filter because $p^{-1}(0)$ is a principal prime ideal.

It turns out that all of these descriptions uniquely determine the initial frame morphism. We sum up:

A point of a locale L is equivalently described as:

- a frame morphism from L to 2
- a principal prime ideal of L
- a meet-prime element of L
- a completely prime filter of L.

All of these descriptions have their place within the theory and it is convenient to switch between them as needed.

The Functor pt

Now that a set of points is available for any locale, it remains to equip this set with an appropriate topology in order to define the object part of the functor pt. This is done by defining the open sets of pt(L) as

$$\varphi(a) = \{ p \mid \text{pt}(L) \mid p(a) = 1 \},$$

for every element a of L. Here we viewed the points of L as morphisms, but one can of course state a similar definition for all of the other equivalent characterisations. It can be shown that setting $\Omega(\text{pt}(L)) = \{\varphi(a) \mid a - L\}$ does really yield a topological space $(\text{pt}(L), \Omega(\text{pt}(L)))$. It is common to abbreviate this space as pt(L).

Finally pt can be defined on morphisms of Frm rather canonically by defining, for a frame morphism g from L to M, $\mathrm{pt}(g)$: $\mathrm{pt}(M) \to \mathrm{pt}(L)$ as $\mathrm{pt}(g)(p) = p \circ g$.

In words, we obtain a morphism from L to 2 (a point of L) by applying the morphism g to get from L to M before applying the morphism p that maps from M to 2. Again, this can be formalized using the other descriptions of points of a locale as well - for example just calculate $(p \circ g)^{-1}(0)$.

The Adjunction of Top and Loc

As noted several times before, pt and Ω usually are not inverses. In general neither is X homeomorphic to $\mathrm{pt}(\Omega(X))$ nor is L order-isomorphic to $\Omega(\mathrm{pt}(L))$.

However, when introducing the topology of $\mathrm{pt}(L)$ above, a mapping φ from L to $\Omega(\mathrm{pt}(L))$ was applied. This mapping is indeed a frame morphism.

Conversely, we can define a continuous function ψ from X to $\mathrm{pt}(\Omega(X))$ by setting $\psi(x) = \Omega(p_x)$, where p_x is just the characteristic function for the point x from 1 to X as described above. Another convenient description is given by viewing points of a locale as meet-prime elements.

In this case we have $\psi(x) = X \setminus \mathrm{Cl}\{x\}$, where $\mathrm{Cl}\{x\}$ denotes the topological closure of the set $\{x\}$ and $\setminus$ is just set-difference.

At this point we already have more than enough data to obtain the desired result: the functors Ω and pt define an adjunction between the categories.

Top and Loc = $\mathrm{Frm}^{\mathrm{op}}$, where pt is right adjoint to Ω and the natural transformations ψ and φ^{op} provide the required unit and counit, respectively.

The Duality Theorem

The above adjunction is not an equivalence of the categories Top and Loc (or, equivalently, a duality of Top and Frm). For this it is necessary that both ψ and φ are isomorphisms in their respective categories. For a space X, ψ: $X \to \mathrm{pt}(\Omega(X))$ is a homeomorphism if and only if it is bijective. Using the characterisation via meet-prime elements of the open set lattice, one sees that this is the case if and only if every meet-prime open set is of the form $X \setminus \mathrm{Cl}\{x\}$ for a unique x.

Alternatively, every join-prime closed set is the closure of a unique point, where "join-prime" can be replaced by (join-) irreducible since we are in a distributive lattice. Spaces with this property are called sober.

Conversely, for a locale L, φ: $L \rightarrow \Omega(\text{pt}(L))$ is always surjective. It is additionally injective if and only if any two elements a and b of L for which a is not less-or-equal to b can be separated by points of the locale, formally:

If not $a \leq b$, then there is a point p in pt(L) such that p(a) = 1 and p(b) = 0.

If this condition is satisfied for all elements of the locale, then the locale is spatial, or said to have enough points.

Finally, one can verify that for every space X, $\Omega(X)$ is spatial and for every locale L, pt(L) is sober. Hence, it follows that the above adjunction of Top and Loc restricts to an equivalence of the full subcategories Sob of sober spaces and SLoc of spatial locales.

This main result is completed by the observation that for the functor pt o Ω, sending each space to the points of its open set lattice is right adjoint to the inclusion functor from Sob to Top. For a space X, pt($\Omega(X)$) is called its soberification.

The case of the functor Ω o pt is symmetric but a special name for this operation is not commonly used.

Specialisation (Pre) Order

In the branch of mathematics known as topology, the specialisation (or canonical) preorder is a natural preorder on the set of the points of a topological space.

For most spaces that are considered in practice, namely for all those that satisfy the T_0 separation axiom, this preorder is even a partial order (called the specialisation order). On the other hand, for T_1 spaces the order becomes trivial and is of little interest.

The specialisation order is often considered in applications in computer science, where T_0 spaces occur in denotational semantics. The specialisation order is also important for identifying suitable topologies on partially ordered sets, as it is done in order theory.

Definition and Motivation

Consider any topological space X. The specialisation preorder $\leq$ on X is defined by setting

$x \leq y$ if and only if cl$\{x\}$ is a subset of cl$\{y\}$,

where cl$\{x\}$ denotes the closure of the singleton set $\{x\}$, i.e. the intersection of all closed sets containing $\{x\}$. While this brief definition is convenient, it is helpful to note that the following statement is equivalent:

$x \leq y$ if and only if y is contained in all open sets that contain x.

This definition explains why one speaks of a "specialisation": y is more special than x, since it is contained in more open sets. This is particularly intuitive if one views open sets as properties that a point x may or may not have.

The more open sets contain a point, the more properties it has, and the more special it is. The usage is consistent with the classical logical notions of genus and species; and also with the traditional use of generic points in algebraic geometry. Specialisation as an idea is applied also in valuation theory.

The intuition of upper elements being more specific is typically found in domain theory, a branch of order theory that has ample applications in computer science.

Upper and Lower Sets

Let X be a topological space and let $\leq$ be the specialisation preorder on X. Every open set is an upper set with respect to $\leq$ and every closed set is a lower set. The converses are not generally true. In fact, a topological space is an Alexandrov space if and only if every upper set is open (or every lower set is closed).

Let A be a subset of X. The smallest upper set containing A is denoted $\downarrow A$ and the smallest lower set containing A is denoted $\downarrow A$. In case $A = \{x\}$ is a singleton one uses the notation $\downarrow x$ and $\downarrow x$. For x - X one has:

- $\uparrow x = \{y \in X : x$ £ $y\} = \cap\{$open sets containing $x\}$.
- $\downarrow x = \{y \in X : y$ £ $x\} = \cap\{$closed sets containing $x\}$ = cl$\{x\}$.

The lower set $\downarrow x$ is always closed; however, the upper set $\downarrow x$ need not be open or closed.

The closed points of a topological space X are precisely the minimal elements of X with respect to $\leq$.

Examples

- In the Sierpinski space {0,1} with open sets {∅ {1}, {0,1 the specialisation order is the natural one ($0 \leq 0$, $0 \leq 1$, and $1 \leq 1$).
- If p, q are elements of Spec(R) (the spectrum of a commutative ring R) then $p \leq q$ if and only if $q \subseteq p$ (as prime ideals). Thus the closed points of Spec(R) are precisely the maximal ideals.

Important Properties

As suggested by the name, the specialisation preorder is a preorder, i.e. it is reflexive and transitive, which is indeed easy to see.

The equivalence relation determined by the specialisation preorder is just that of topological indistinguishability. That is, x and y are topologically indistinguishable if and only if $x \leq y$ and $y \leq x$. Therefore, the antisymmetry of $\leq$ is precisely the T_0 separation axiom: if x and y are indistinguishable then $x = y$. In this case it is justified to speak of the specialisation order.

On the other hand, the symmetry of specialisation preorder is equivalent to the R_0 separation axiom: $x \leq y$ if and only if x and y are topologically indistinguishable.

It follows that if the underlying topology is T_1, then the specialisation order is discrete, i.e. one has $x \leq y$ if and only if $x = y$. Hence, the specialisation order is of little interest for T_1 topologies, especially for all Hausdorff spaces.

Any continuous function between two topological spaces is monotone with respect to the specialisation preorders of these spaces. The converse, however, is not true in general.

In the language of category theory, we then have a functor from the category of topological spaces to the category of preordered sets which assigns a topological space its specialisation preorder. This functor has a left adjoint which places the Alexandrov topology on a preordered set.

There are spaces that are more specific than T_0 spaces for which this order is interesting: the sober spaces. Their relationship to the specialisation order is more subtle:

For any sober space X with specialisation order $\leq$, we have

- $(X, \leq)$ is a directed complete partial order, i.e. every directed subset S of $(X, \leq)$ has a supremum sup S,
- for every directed subset S of $(X, \leq)$ and every open set O, if sup S is in O, then S and O have non-empty intersection.

One may describe the second property by saying that open sets are *inaccessible by directed suprema*. A topology is order consistent with respect to a certain order $\leq$ if it induces $\leq$ as its specialisation order and it has the above property of inaccessibility with respect to (existing) suprema of directed sets in $\leq$.

Topologies on Orders

The specialisation order yields a tool to obtain a partial order from every topology. It is natural to ask for the converse too: Is every partial order obtained as a specialisation order of some topology?

Indeed, the answer to this question is positive and there are in general many topologies on a set X which induce a given order $\leq$ as their specialisation order.

The Alexandroff topology of the order $\leq$ plays a special role: it is the finest topology that induces $\leq$. The other extreme, the coarsest topology that induces $\leq$, is the upper topology, the least topology within which all complements of sets $\{y \text{ in } X \mid y \leq x\}$ (for some x in X) are open.

There are also interesting topologies in between these two extremes. The finest topology that is order consistent in the above sense for a given order $\leq$ is the Scott topology.

The upper topology however is still the coarsest order consistent topology. In fact its open sets are even inaccessible by *any* suprema. Hence any sober space with specialisation order $\leq$ is finer than the upper topology and coarser than the Scott topology. Yet, such a space may fail to exist. Especially, the Scott topology is not necessarily sober.

Sober Space

In mathematics, a sober space is a topological space such that every irreducible closed subset of *X* is the closure of exactly one point of *X*: that is, this closed subset has a unique generic point.

Properties and Examples

Any Hausdorff (T_2) space is sober (the only irreducible subsets being points), and all sober spaces are Kolmogorov (T_0), and both implications are strict. Sobriety is not comparable to the T_1 condition: an example of a T_1 space which is not sober is an infinite set with the cofinite topology.

Moreover T_2 is stronger than T_1 *and* sober, i.e., while every T_2 space is at once T_1 and sober, there exist spaces that are simultaneously T_1 and sober, but not T_2.

One such example is the following: let X be the set of real numbers, with a new point p adjoined; the open sets being all real open sets, and all cofinite sets containing p.

Sobriety of *X* is precisely a condition that forces the lattice of open subsets of *X* to determine *X* up to homeomorphism, which is relevant to pointless topology. Sobriety makes the specialisation preorder a directed complete partial order.

The prime spectrum Spec(*R*) of a commutative ring *R* with the Zariski topology is a compact sober T_0 space. In fact, every spectral space (i.e. a compact sober space for which the collection of compact open subsets is closed under finite intersections and forms a base for the topology) is homeomorphic to Spec(*R*) for some commutative ring *R*. This is a theorem of Melvin Hochster.

More generally, the underlying topological space of any scheme is a sober space.

Spectral Space

In mathematics, a spectral space is a topological space which is homeomorphic to the spectrum of a commutative ring.

Definition

Let *X* be a topological space and let $K\circ(X)$ be the set of all quasi-compact and open subsets of *X*. Then *X* is said to be *spectral* if it satisfies all of the following conditions:

- X is quasi-compact and T_0.
- $K_o(X)$ is a basis of open subsets of X.
- $K_o(X)$ is closed under finite intersections.
- X is sober, i.e. every nonempty irreducible closed subset of X has a (necessarily unique) generic point.

Equivalent Descriptions

Let X be a topological space. Each of the following properties are equivalent to the property of X being spectral:

1. X is homeomorphic to a projective limit of finite T_0-spaces.
2. X is homeomorphic to the spectrum of a bounded distributive lattice L. In this case, L is isomorphic (as a bounded lattice) to the lattice $K(X)$ (this is called Stone representation of distributive lattices).
3. X is homeomorphic to the spectrum of a commutative ring.
4. X is the topological space determined by a Priestley space.
5. X is a coherent space in the sense of topology (this indeed is only another name).

Properties

Let X be a spectral space and let $K(X)$ be as before. Then:

- $K_o(X)$ is a bounded sublattice of subsets of X.
- Every closed subspace of X is spectral.
- An arbitrary intersection of quasi-compact and open subsets of X (hence of elements from $K_o(X)$) is again spectral.
- X is T_0 by definition, but in general not T_1. In fact a spectral space is T_1 if an only if it is Hausdorff (or T_2) if and only if it is a boolean space.
- X can be seen as a Pairwise Stone space.

Spectral Maps

A spectral map $f: X \to Y$ between spectral spaces X and Y is a continuous map such that the preimage of every open and quasi-compact subset of Y under f is again quasi-compact.

The category of spectral spaces which has spectral maps as morphisms is dually equivalent to the category of bounded distributive lattices (together with morphisms of such lattices. In this anti-equivalence, a spectral space X corresponds to the lattice K(X).

Alexandrov Topology

In topology, an Alexandrov space (or Alexandrov-discrete space) is a topological space in which the intersection of any family of open sets is open. It is an axiom of topology that the intersection of any *finite* family of open sets is open. In an Alexandrov space the finite restriction is dropped.

Alexandrov topologies are uniquely determined by their specialisation preorders. Indeed, given any preorder $\leq$ on a set X, there is a unique Alexandrov topology on X for which the specialisation preorder is $\leq$. The open sets are just the upper sets with respect to $\leq$. Thus, Alexandrov topologies on X are in one-to-one correspondence with preorders on X.

Alexandrov spaces are also called finitely generated spaces since their topology is uniquely determined by the family of all finite subspaces. Alexandrov spaces can be viewed as a generalization of finite topological spaces.

Characterisations of Alexandrov Topologies

Alexandrov topologies have numerous characterisations. Let $X = \langle X, T \rangle$ be a topological space. Then the following are equivalent:

- Open and closed set characterisations:
  - Open set characterisation. An arbitrary intersection of open sets in X is open.
  - Closed set characterisation. An arbitrary union of closed sets in X is closed.
- Neighbourhood characterisations:
  - Smallest neighbourhood characterisation. Every point of X has a smallest neighbourhood.
  - Neighbourhood filter characterisation. The neighbourhood filter of every point in X is closed under arbitrary intersections.

- Interior and closure algebraic characterisations:
  - Interior operator characterisation. The interior operator of X distributes over arbitrary intersections of subsets.
  - Closure operator characterisation. The closure operator of X distributes over arbitrary unions of subsets.
- Preorder characterisations:
  - Specialisation preorder characterisation. T is the finest topology consistent with the specialisation preorder of X i.e. the finest topology giving the preorder $\leq$ satisfying $x \leq y$ if and only if x is in the closure of $\{y\}$ in X.
  - Open up-set characterisation. There is a preorder $\rightarrow$ such that the open sets of X are precisely those that are upwardly closed i.e. if x is in the set and $x \leq y$ then y is in the set. (This preorder will be precisely the specialisation preorder.)
  - Closed down-set characterisation. There is a preorder $\rightarrow$ such that the closed sets of X are precisely those that are downwardly closed i.e. if x is in the set and y £ x then y is in the set. (This preorder will be precisely the specialisation preorder.)
  - Upward interior characterisation. A point x lies in the interior of a subset S of X if and only if there is a point y in S such that $y \leq x$ where $\leq$ is the specialisation preorder i.e. y lies in the closure of $\{x\}$.
  - Downward closure characterisation. A point x lies in the closure of a subset S of X if and only if there is a point y in S such that $x \leq y$ where $\leq$ is the specialisation preorder i.e. x lies in the closure of $\{y\}$.
- Finite generation and category theoretic characterisations:
  - Finite closure characterisation. A point x lies within the closure of a subset S of X if and only if there is a finite subset F of S such that x lies in the closure of F.

- Finite subspace characterisation. T is coherent with the finite subspaces of X.
- Finite inclusion map characterisation. The inclusion maps $f_i : X_i \to X$ of the finite subspaces of X form a final sink.
- Finite generation characterisation. X is finitely generated i.e. it is in the final hull of the finite spaces. (This means that there is a final sink $f_i : X_i \to X$ where each X_i is a finite topological space.)

Topological spaces satisfying the above equivalent characterisations are called finitely generated spaces or Alexandrov spaces and their topology T is called the Alexandrov topology, named after the Russian mathematician Pavel Alexandrov who first investigated them.

Equivalence between Monotony and Continuity

Given a monotone function

$$f : X \to Y$$

between two preordered sets (i.e. a function

$$f : X \to Y$$

between the underlying sets such that $x \leq y$ in X implies $f(x) \leq f(y)$ in Y), let

$$T(f) : T(X) \to T(Y)$$

be the same map as f considered as a map between the corresponding Alexandrov spaces. Then

$$T(f) : T(X) \to T(Y)$$

is a continuous map.

Conversely given a continuous map

$$f : X \to Y$$

between two topological spaces, let

$$W(f) : W(X) \to W(Y)$$

be the same map as f considered as a map between the corresponding preordered sets. Then

$$W(f) : W(X) \to W(Y)$$

is a monotone function.

Thus a map between two preordered sets is monotone if and only if it is a continuous map between the corresponding Alexandrov spaces. Conversely a map between two Alexandrov spaces is continuous if and only if it is a monotone function between the corresponding preordered sets.

Notice however that in the case of topologies other than the Alexandrov topology, we can have a map between two topological spaces that is not continuous but which is nevertheless still a monotone function between the corresponding preordered sets.

$$i : X \to T(W(X)).)$$

Category Theoretic Description of the Duality

Let Set denote the category of sets and maps. Let Top denote the category of topological spaces and continuous maps; and let Pro denote the category of preordered sets and monotone functions. Then

$$T : \text{Pro} \to \text{Top and}$$

$$W : \text{Top} \to \text{Pro}$$

are concrete functors over Set which are left and right adjoints respectively.

Let Alx denote the full subcategory of Top consisting of the Alexandrov spaces. Then the restrictions

$$T : \text{Pro} \to \text{Alx and}$$

$$W : \text{Alx} \to \text{Pro}$$

are inverse concrete isomorphisms over Set.

Alx is in fact a coreflective subcategory of Top with coreflector $T\zeta\%W : \text{Top} \to \text{Alx}$. This means that given a topological space X, the identity map

$$i : T(W(X)) \to X$$

is continuous and for every continuous map

$$f : Y \to X$$

where Y is an Alexandrov space, the composition

$$i^{-1}\zeta\%f : Y \text{-}\!\!\to T(W(X))$$

is continuous.

Relationship to the Construction of Modal Algebras from Modal Frames

Given a preordered set X, the interior operator and closure operator of $T(X)$ are given by:

$\mathrm{Int}(S) = \{ x \in X : \text{for all } y \in X, x \leq y \text{ implies } y \in S \}$, for all $S \subseteq X$

$\mathrm{Cl}(S) = \{ x \in X : \text{there exists a } y \in S \text{ with } x \leq y \}$ for all $S \subseteq X$

Considering the interior operator and closure operator to be modal operators on the power set Boolean algebra of X, this construction is a special case of the construction of a modal algebra from a modal frame i.e. a set with a single binary relation. (The latter construction is itself a special case of a more general construction of a complex algebra from a relational structure i.e. a set with relations defined on it.) The class of modal algebras that we obtain in the case of a preordered set is the class of interior algebras—the algebraic abstractions of topological spaces.

History

Alexandrov spaces were first introduced in 1937 by P. S. Alexandrov under the name discrete spaces, where he provided the characterisations in terms of sets and neighbourhoods. The name discrete spaces later came to be used for topological spaces in which every subset is open and the original concept lay forgotten. With the advancement of categorical topology in the 1980s, Alexandrov spaces were rediscovered when the concept of finite generation was applied to general topology and the name finitely generated spaces was adopted for them. Alexandrov spaces were also rediscovered around the same time in the context of topologies resulting from denotational semantics and domain theory in computer science.

In 1966 Michael C. McCord and A. K. Steiner each independently observed a duality between partially ordered sets and spaces which were precisely the T_0 versions of the spaces that Alexandrov had introduced. P. Johnstone referred to such topologies as Alexandrov topologies. F. G. Arenas independently proposed this name for the general version of these topologies. McCord also showed that these spaces are weak homotopy equivalent to the order complex of the corresponding partially ordered set. Steiner demonstrated that the duality is a contravariant lattice isomorphism preserving arbitrary meets and joins as well as complementation.

It was also a well known result in the field of modal logic that a duality exists between finite topological spaces and preorders on finite sets (the finite modal frames for the modal logic *S4*). C. Naturman extended these results to a duality between Alexandrov spaces and preorders in general, providing the preorder characterisations as well as the interior and closure algebraic characterisations.

A systematic investigation of these spaces from the point of view of general topology which had been neglected since the original paper by Alexandrov, was taken up by F.G. Arenas.

Inspired by the use of Alexandrov topologies in computer science, applied mathematicians and physicists in the late 1990s began investigating the Alexandrov topology corresponding to causal sets which arise from a preorder defined on spacetime modeling causality.

Nerve of a Covering

In mathematics, the nerve of an open covering is a construction in topology, of an abstract simplicial complex from an open covering of a topological space X.

The notion of nerve was introduced by Pavel Alexandrov.

Given an index set I, and open sets U_i contained in X, the *nerve* N is the set of finite subsets of I defined as follows:

- the empty set belongs to N;
- a finite set $J \subseteq I$ belongs to N if and only if the intersection of the U_i whose subindices are in J is non-empty. That is, if and only if

$$\bigcap_{j \in J} U_j \neq \varnothing.$$

Obviously, if J belongs to N, then any of its subsets is also in N. Therefore N is an abstract simplicial complex.

In general, the complex N need not reflect the topology of X accurately. For example we can cover any n-sphere with two contractible sets U and V, in such a way that N is a 1-simplex. However, if we also insist that the open sets corresponding to every intersection indexed by a set in N is also contractible, the position changes. This means for instance that a circle covered by three open arcs, intersecting in pairs in one arc, is modelled by a homeomorphic complex, the geometrical realisation of N

Lawson Topology

In mathematics and theoretical computer science the Lawson topology, named after J. D. Lawson, is a topology on partially ordered sets used in the study of domain theory. The lower topology on a poset P is generated by the subbasis consisting of all complements of principal filters on P. The Lawson topology on P is the smallest common refinement of the lower topology and the Scott topology on P.

Properties

- If P is a complete upper semilattice, the Lawson topology on P is always a complete T_1 topology.

Inductive Dimension

In the mathematical field of topology, the inductive dimension of a topological space X is either of two values, the small inductive dimension ind(X) or the large inductive dimension Ind(X). These are based on the observation that, in n-dimensional Euclidean space R^n, (n - 1)-dimensional spheres (that is, the boundaries of n-dimensional balls) have dimension n - 1. Therefore it should be possible to define the dimension of a space inductively in terms of the dimensions of the boundaries of suitable open sets.

The small and large inductive dimensions are two of the three most usual ways of capturing the notion of "dimension" for a topological space, in a way that depends only on the topology (and not, say, on the properties of a metric space). The other is the Lebesgue covering dimension. The term "topological dimension" is ordinarily understood to refer to Lebesgue covering dimension. For "sufficiently nice" spaces, the three measures of dimension are equal.

Formal Definition

We want the dimension of a point to be 0, and a point has empty boundary, so we start with

$$\text{ind}(\varnothing) = \text{Ind}(\varnothing) = -1$$

Then inductively, ind(X) is the smallest n such that, for every $x \in X$ and every open set U containing x, there is an open V containing x, where the closure of V is a subset of U, such that the boundary of V has small inductive dimension less than or equal to n - 1. (In the case above, where X is Euclidean n-dimensional space, V will be chosen to be an n-dimensional ball centred at x.)

For the large inductive dimension, we restrict the choice of V still further; Ind(X) is the smallest n such that, for every closed subset F of every open subset U of X, there is an open V in between (that is, F is a subset of V and the closure of V is a subset of U), such that the boundary of V has large inductive dimension less than or equal to n - 1.

Relationship between Dimensions

Let $\dim$ be the Lebesgue covering dimension. For any topological space X, we have

$$\dim X = 0 \text{ if and only if } \text{Ind}\, X = 0.$$

Urysohn's theorem states that when X is a normal space with a countable base, then

$$\dim X = \text{Ind}\, X = \text{ind}\, X.\cdot$$

Such spaces are exactly the separable and metrizable X.

The Nöbeling-Pontryagin theorem then states that such spaces with finite dimension are characterised up to homeomorphism as the subspaces of the Euclidean spaces, with their usual topology. The Menger-Nöbeling theorem (1932) states that if X is compact metric separable and of dimension n, then it embeds as a subspace of Euclidean space of dimension $2n$ + 1. (Georg Nöbeling was a student of Karl Menger. He introduced Nöbeling space, the subspace of R^{2n+1} consisting of points with at least n + 1 co-ordinates being irrational numbers, which has universal properties for embedding spaces of dimension n.)

Assuming only X metrizable we have (Miroslav Katìtov)

$$\text{ind}\ X \leq \text{Ind}\ X = \dim X;$$

or assuming X compact and Hausdorff (P. S. Aleksandrov)

$$\dim X \leq \text{ind}\ X \leq \text{Ind}\ X.$$

Either inequality here may be strict; an example of Vladimir V. Filippov shows that the two inductive dimensions may differ.

A separable metric space X satisfies the inequality $IndX \leq n$ if and only if for every closed sub-space A of the space X and each continuous mapping $f: A \rightarrow S^n$ there exists a continuous extension $\overline{f}: X \rightarrow S^n$..

Lebesgue Covering Dimension

Lebesgue covering dimension or topological dimension is one of several inequivalent notions of assigning a topological invariant dimension to a given topological space.

Definition

The covering dimension of a topological space X is defined to be the minimum value A of n, such that every finite open cover A of X admits a finite open cover of X which refines A in which no point is included in more than n+1 elements. If no such minimal n exists, the space is said to be of infinite covering dimension.

Examples

The n-dimensional Euclidean space $\mathbb{E}^n$ has covering dimension n. A topological space is zero-dimensional with respect to the covering dimension if every open cover of the space has a refinement consisting of disjoint open sets so that any point in the space is contained in exactly one open set of this refinement.

Any given open cover of the unit circle will have a refinement consisting of a collection of open arcs. The circle has dimension one, by this definition, because any such cover can be further refined to the stage where a given point x of the circle is contained in *at most* two open arcs. That is, whatever collection of arcs we begin with, some can be discarded or shrunk, such that the remainder still covers the circle, but with simple overlaps.

Similarly, any open cover of the unit disk in the two-dimensional plane can be refined so that any point of the disk is contained in no more than three open sets, while two are in general not sufficient. The covering dimension of the disk is thus two.

A non-technical illustration of these examples below.

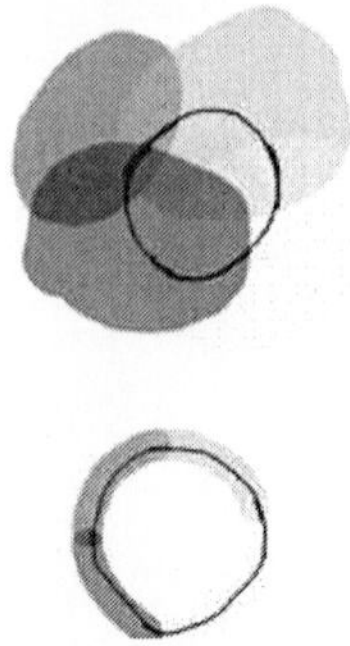

Figure: *Below is a refinement of a cover (above) of a circular line (black). Notice how in the refinement no point on the line is contained in more than two sets. Note also how the sets link to each other to form a "chain".*

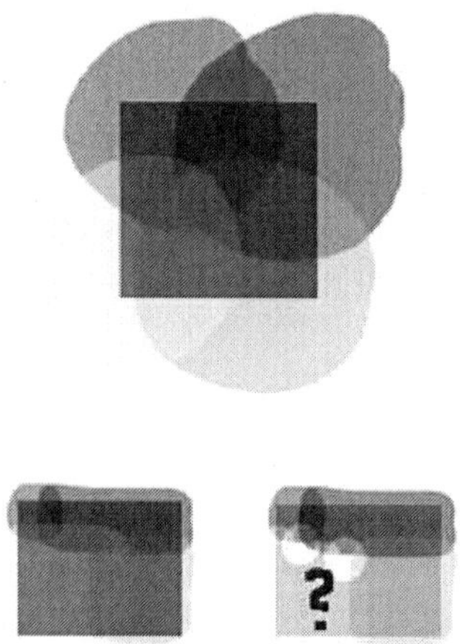

Figure: *Below left is a refinement of a cover (above) of a planar shape (dark) so that all points in the shape are contained in at most three sets. Below right is an attempt to refine the cover so that no point would be contained in more than two sets. This fails in the intersection of set borders. Thus, a planar shape isn't "webby" or cannot be covered with "chains", but is in a sense thicker; i.e., its topological dimension must be higher than one.*

Lebesgue's Number Lemma

In topology, Lebesgue's number lemma, named after Henri Lebesgue, is a useful tool in the study of compact metric spaces. It states:

If the metric space *(X, d)* is compact and an open cover of *X* is given, then there exists a number $\delta > 0$ such that every subset of *X* having diametre less than ä is contained in some member of the cover.

Such a number δ is called a Lebesgue number of this cover. The notion of a Lebesgue number itself is useful in other applications as well.

Barycentric Subdivision

In geometry, the barycentric subdivision is a standard way of dividing an arbitrary convex polygon into triangles, a convex polyhedron into tetrahedra, or, in general, a convex polytope into simplices with the same dimension, by connecting the barycentres of their faces in a specific way.

The name is also used in topology for a similar operation on cell complexes. The result is topologically equivalent to that of the geometric operation, but the parts have arbitrary shape and size. Both operations have a number of applications in mathematics and in geometric modeling, especially whenever some function or shape needs to be approximated piecewise, e.g. by a spline.

Barycentric Subdivision in Topology

Barycentric subdivision is an important tool in simplicial homology theory, where it is used as a means of obtaining finer simplicial complexes (containing the original ones, i.e. with more simplices). This in turn is crucial to the simplicial approximation theorem, which roughly states that one can approximate any continuous function between polyhedra by a (finite) simplicial map, given a sufficient amount of subdivision of the respective simplicial complexes whom they realise. Ultimately, this approximation technique is a standard ingredient in the proof that simplicial homology groups are topological invariants.

A generalization of barycentric subdivision can also be defined for a cell complex. Informally, such an object can be thought of as an assemblage of one or more chunks of rubber (*cells*), each shaped like a convex polytope, which are glued to each other by their facets — possibly with much stretching and twisting.

The topological version of BCS replaces each cell by an assemblage of rubber simplices, likewise glued together by their facets and possibly deformed. The procedure is (1) select for each cell a deformation map that converts it into a geometric convex polytope, preserving its incidence and topological connections; (2) perform the geometric BCS on this polytope; and, then (3) map the resulting subdivision back to the original cells.

The result of barycentric subdivision, when viewed as an abstract simplicial complex, is an example of a flag complex. It has one vertex for every cell of the original cell complex and one maximal-dimensional cell for every *flag* (a collection of cells of different dimensions, all related to each other by inclusion) of the original cell complex.

Applications

The barycentric subdivision is chiefly used to replace an arbitrarily complicated convex polytope or topological cell complex by an assemblage of pieces, all of them of bounded complexity (simplices, in fact). A typical application is modeling the shape of a car body by a spline — a piecewise-defined polynomial function. The algebra of such functions becomes much simpler and easier to program if each "piece" is a "topological triangle", i.e. is attached to exactly three other pieces. However, a human user may find it more natural to design the shape by joining patches with more liberal shapes and topologies. Barycentric subdivision is a convenient way to convert that "user-friendly" model into a "computer-friendly" one.

Repeated Barycentric Subdivision

When approximating a mathematical function or a surface by a spline, the accuracy of the approximation is usually determined by the piece size — the bigger the pieces, the larger the error. Thus it is often necessary to split large pieces into smaller ones, in order to achieve a prescribed accuracy.

In theory, BCS could be used for that purpose, since it has the property that the longest edge of any piece is smaller than the longest edge of the original polytope by a factor less than $n/(n+1)$.. Therefore, by applying BCS sufficiently many times, the largest edge can be made as small as desired.

However, in practice BCS is not well-suited for that purpose. For one thing, each application after the first one multiplies the number of simplices by $(n-1)!$.. BCS also multiplies the degree of each original vertex by n!, and the degree of each edge by $(n+1)!$.

Moreover, the BCS will split all simplices, even those that are already small enough. Finally, each BCS stage also makes the simplices not only smaller but "skinnier", i.e. it tends to increase their *aspect ratio* (the ratio between the longest and shortest edge). For all these reasons, in practice one rarely applies more than one round of BCS, and other subdivision schemes are used instead.

Relative Barycentric Subdivision

For simplicial complexes $L \subset K$ one defines the relative barycentric subdivision K^* of K *modulo L* that consists of those simplixes with vertices $v_0 \ldots v_k B(S_1') \ldots B(S_l')$ associated to a sequence $S_0 < \cdots < S_k$ of proper faces of L and barycentres $B(S_i')$ of simplexes in $K \setminus L$.

Clearly, L remains a subcomplex of K^*. Only the simplexes away from L shrink.

Related Notions

False barycentric subdivision: Sometimes the term "barycentric subdivision" is improperly used for any subdivision of a polytope P into simplices that have one vertex at the centroid of P, and the opposite facet on the boundary of P. While this property holds for the true barycentric subdivision, it also holds for other subdivisions which are not the BCS.

For example, if one makes a straight cut from the barycentre of a triangle to each of its three corners, one obtains a subdivision into three triangles. Generalizing this idea, one obtains a schema for subdividing an n-dimensional simplex into n+1 simplices. However, this subdivision is not the BCS.

Simplicial Sets

The barycentric division can also be defined for simplicial sets, in a way that is compatible (with respect to the topological realisation functor) with the above division of simplices.

In mathematics, a simplicial set is a construction in categorical homotopy theory which is a purely algebraic model of the notion of a "well-behaved" topological space. Historically, this model arose from earlier work in combinatorial topology and in particular from the notion of simplicial complexes.

Motivation

A simplicial set is a categorical (that is, purely algebraic) model capturing those topological spaces which can be built up (or faithfully represented up to homotopy) from simplices and their incidence relations. This is similar to the approach of CW complexes to modeling topological spaces, with the crucial difference that simplicial sets are purely algebraic and do not carry any actual topology (this will become clear in the formal definition).

To get back to actual topological spaces, there is a *geometric realisation* functor available which takes values in the category of compactly generated Hausdorff spaces.

Most classical results on CW complexes in homotopy theory have analogous versions for simplicial sets which generalize these results. While algebraic topologists largely continue to prefer CW complexes, there is a growing contingent of researchers interested in using simplicial sets for applications in algebraic geometry where CW complexes do not naturally exist.

Formal Definition

Using the language of category theory, a simplicial set X is a contravariant functor

$$X\colon \Delta \to \text{Set}$$

where Δ denotes the simplex category whose objects are finite strings of ordinal numbers of the form

$$n = 0 \to 1 \to \ldots \to n$$

(or in other words non-empty totally ordered finite sets) and whose morphisms are order-preserving functions between them, and Set is the category of small sets. It is common to define simplicial sets as a covariant functor from the opposite category, as

$$X\colon \Delta^{\text{op}} \to \text{Set}$$

This definition is clearly equivalent to the one immediately above.

Alternatively, one can think of a simplicial set as a simplicial object in the category Set, but this is only different language for the definition just given. If we use a *covariant* functor X instead of a contravariant one, we arrive at the definition of a cosimplicial set.

Simplicial sets form a category usually denoted sSet or just S whose objects are simplicial sets and whose morphisms are natural transformations between them. There is a corresponding category for cosimplicial sets as well, denoted by cSet.

These definitions arise from the relationship of the conditions imposed on the *face maps* and *degeneracy maps* to the category Δ.

Face and Degeneracy Maps

In Δ^{op}, there are two particularly important classes of maps called face maps and degeneracy maps which capture the underlying combinatorial structure of simplicial sets.

The face maps $d_i : n \to n - 1$ are given by

$d_i\,(0 \to \dots \to n) = (0 \to \dots \to i - 1 \to i + 1 \to \dots \to n)$.

The degeneracy maps $s_i : n \to n + 1$ are given by

$s_i\,(0 \to \dots \to n) = (0 \to \dots \to i - 1 \to i \to i \to i + 1 \to \dots \to n)$.

By definition, these maps satisfy the following simplicial identities:

$d_i\, d_j = d_{j-1}\, d_i$ if $i < j$

$d_i\, s_j = s_{j-1}\, d_i$ if $i < j$

$d_j\, s_j = \mathrm{id} = d_{j+1}\, s_j$

$d_i\, s_j = s_j\, d_{i-1}$ if $i > j + 1$

$s_i\, s_j = s_{j+1}\, s_i$ if $i \leq j$.

The simplicial category Δ has as its morphisms the monotonic non-decreasing functions. Since the morphisms are generated by those that 'skip' or 'add' a single element, the detailed relations written out above underlie the topological applications. It can be shown that these relations suffice.

Chain (Algebraic Topology)

In algebraic topology, a simplicial k-chain is a formal linear combination of k-simplices.

Integration on Chains

Integration is defined on chains by taking the linear combination of integrals over the simplices in the chain with coefficients typically integers. The set of all k-chains forms a group and the sequence of these groups is called a chain complex.

Boundary Operator on Chains

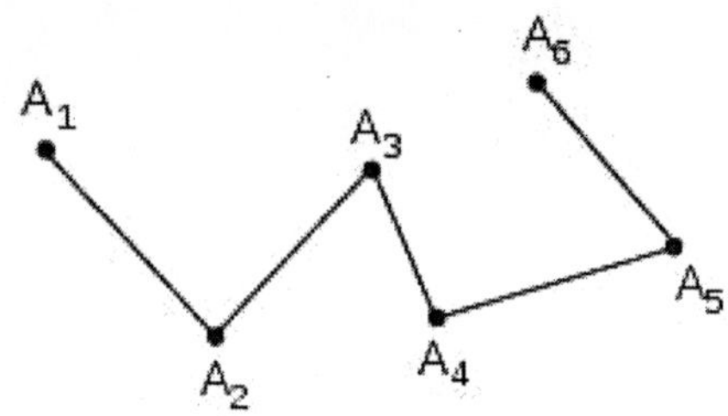

Figure: *The boundary of a polygonal curve is a linear combination of its nodes; in this case, some linear combination of* A_1 *through* A_6. *Assuming the segments all are oriented left-to-right (in increasing order from* A_k *to* A_{k+1}*), the boundary is* $A_6 - A_1$.

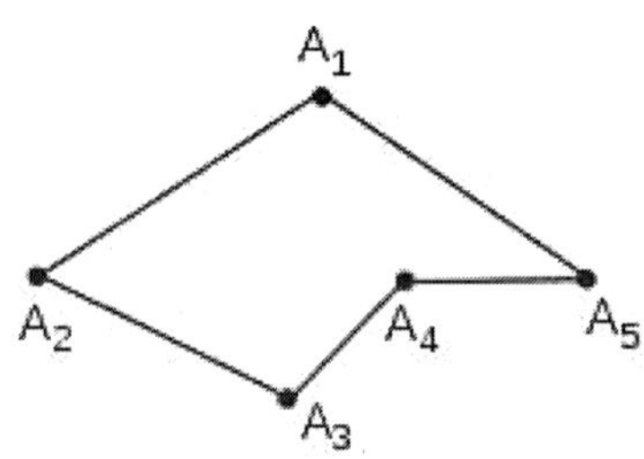

Figure: *A closed polygonal curve, assuming consistent orientation, has null boundary.*

The boundary of a chain is the linear combination of boundaries of the simplices in the chain. The boundary of a k-chain is a (k-1)-chain. Note that the boundary of a simplex is not a simplex, but a chain with coefficients 1 or -1 – thus chains are the closure of simplices under the boundary operator.

Example: The boundary of a path is the formal difference of its endpoints: it is a telescoping sum. To illustrate, if the 1-chain

$c = t_1 + t_2 + t_3$ is a path from point v_1 to point v_4, where $t_1 = [v_1, v_2], t_2 = [v_2, v_3]$, and $t_3 = [v_3, v_4]$ are its constituent 1-simplices, then

$$\begin{aligned}\partial_1 c &= \partial_1(t_1 + t_2 + t_3) \\ &= \partial_1(t_1) + \partial_1(t_2) + \partial_1(t_3) \\ &= \partial_1([v_1, v_2]) + \partial_1([v_2, v_3]) + \partial_1([v_3, v_4]) \\ &= ([v_2] - [v_1]) + ([v_3] - [v_2]) + ([v_4] - [v_3]) \\ &= [v_4] - [v_1].\end{aligned}$$

Example: The boundary of the triangle is a formal sum of its edges with signs arranged to make the traversal of the boundary counterclockwise.

A chain is called a cycle when its boundary is zero. A chain that is the boundary of another chain is called a boundary. Boundaries are cycles, so chains form a chain complex, whose homology groups (cycles modulo boundaries) are called simplicial homology groups.

Example: A 0-cycle is a linear combination of points such that the sum of all the coefficients is 0. Thus, the 0-homology group measures the number of path connected components of the space.

Example: The plane punctured at the origin has nontrivial 1-homology group since the unit circle is a cycle, but not a boundary.

In differential geometry, the duality between the boundary operator on chains and the exterior derivative is expressed by the general Stokes' theorem.

Betti Number

In algebraic topology, a mathematical discipline, the Betti numbers can be used to distinguish topological spaces. Intuitively, the first Betti number of a space counts the maximum number of cuts that can be made without dividing the space into two pieces.

Each Betti number is a natural number or +". For the most reasonable finite-dimensional spaces (such as compact manifolds, finite simplicial complexes or CW complexes), the sequence of Betti numbers is 0 from some points onwards (Betti numbers vanish above the dimension of a space), and they are all finite.

The term "Betti numbers" was coined by Henri Poincaré after Enrico Betti.

Informal Definition

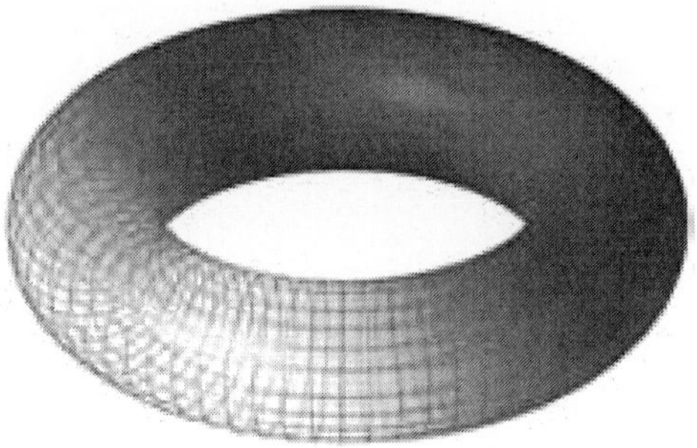

Figure: *A torus has one connected component, two circular holes (the one in the centre and the one in the middle of the "tube"), and one three-dimensional void (the inside of the "tube") yielding Betti numbers of 1,2,1.*

Informally, the kth Betti number refers to the number of unconnected k-dimensional surfaces. The first few Betti numbers have the following intuitive definitions:

- b_0 is the number of connected components
- b_1 is the number of two-dimensional or "circular" holes
- b_2 is the number of three-dimensional holes or "voids"

Definition

For a non-negative integer k, the kth Betti number $b_k(X)$ of the space X is defined as the rank of the abelian group $H_k(X)$, the kth homology group of X. Equivalently, one can define it as the vector space dimension of $H_k(X; Q)$, since the homology group in this case is a vector space over Q. The universal coefficient theorem, in a very simple case, shows that these definitions are the same.

More generally, given a field F one can define $b_k(X, F)$, the kth Betti number with coefficients in F, as the vector space dimension of $H_k(X, F)$.

Example: the first Betti number in graph theory

In topological graph theory the first Betti number of a graph G with n vertices, m edges and k connected components equals

$$m-n+k.$$

This may be proved straightforwardly by mathematical induction on the number of edges. A new edge either increments the number of 1-cycles or decrements the number of connected components.

Properties

The (rational) Betti numbers $b_k(X)$ do not take into account any torsion in the homology groups, but they are very useful basic topological invariants. In the most intuitive terms, they allow one to count the number of *holes* of different dimensions. For a circle, the first Betti number is 1. For a general pretzel, the first Betti number is twice the number of holes.

In the case of a finite simplicial complex the homology groups $H_k(X, Z)$ are finitely-generated, and so has a finite rank. Also the group is 0 when k exceeds the top dimension of a simplex of X.

For a finite CW-complex K we have

$$\chi(K)=\sum_{i=0}^{\infty}(-1)^i b_i(K,F),$$

where $\chi(K)$ denotes Euler characteristic of K and any field F.

For any two spaces X and Y we have

$$P_{X\times Y}=P_X P_Y,$$

where P_X donotes the Poincaré polynomial of X, (more generally, the Poincaré series, for infinite-dimensional spaces), i.e. the generating function of the Betti numbers of X:

$$P_X(z)=b_0(X)+b_1(X)z+b_2(X)z^2+\cdots,$$

If X is n-dimensional manifold, there is symmetry interchanging k and n - k, for any k:

$$b_k(X)=b_{n-k}(X),$$

under conditions (a *closed* and *oriented* manifold).

The dependence on the field F is only through its characteristic. If the homology groups are torsion-free, the

Betti numbers are independent of F. The connection of p-torsion and the Betti number for characteristic p, for p a prime number, is given in detail by the universal coefficient theorem (based on Tor functors, but in a simple case).

Examples:

1. The Betti number sequence for a circle is 1, 1, 0, 0, 0, ...;
   the Poincaré polynomial is
   $$1+x..$$
2. The Betti number sequence for a two-torus is 1, 2, 1, 0, 0, 0, ...;
   the Poincaré polynomial is
   $$(1+x)^2 = 1+2x+x^2..$$
3. The Betti number sequence for a three-torus is 1, 3, 3, 1, 0, 0, 0, ...
   the Poincaré polynomial is
   $$(1+x)^3 = 1+3x+3x^2+x^3..$$
4. Similarly, for an n-torus,

the Poincaré polynomial is

$(1+x)^n$ (by the Künneth theorem), so the Betti numbers are the binomial coefficients.

It is possible for spaces that are infinite-dimensional in an essential way to have an infinite sequence of non-zero Betti numbers. An example is the infinite-dimensional complex projective space, with sequence 1, 0, 1, 0, 1, ... that is periodic, with period length 2. In this case the Poincaré function is not a polynomial but rather an infinite series

$$1+x^2+x^4+\cdots,,$$

which, being a geometric series, can be expressed as the rational function

$$\frac{1}{1-x^2} = 1+x^2+(x^2)^2+(x^2)^3+\cdots.$$

More generally, any sequence that is periodic can be expressed as a sum of geometric series, generalizing the above (e.g. $a,b,c,a,b,c,\ldots,$, has generating function

$(a+bx+cx^2)/(1-x^3))$,

and more generally linear recursive sequences are exactly the sequences generated by rational functions; thus the Poincaré series is expressible as a rational function if and only if the sequence of Betti numbers is a linear recursive sequence.

The Poincaré polynomials of the compact simple Lie groups are:

$$P_{SU(n+1)}(x) = (1+x^3)(1+x^5)...(1+x^{2n+1})$$

$$P_{SO(2n+1)}(x) = (1+x^3)(1+x^7)...(1+x^{4n+1})$$

$$P_{Sp(n)}(x) = (1+x^3)(1+x^7)...(1+x^{4n+1})$$

$$P_{SO(2n)}(x) = (1+x^3)(1+x^7)...(1+x^{2n-1})(1+x^{4n-5})$$

$$P_{G_2}(x) = (1+x^3)(1+x^{11})$$

$$P_{F_4}(x) = (1+x^3)(1+x^{11})(1+x^{15})(1+x^{23})$$

$$P_{E_6}(x) = (1+x^3)(1+x^9)(1+x^{11})(1+x^{15})(1+x^{17})(1+x^{23})$$

$$P_{E_7}(x) = (1+x^3)(1+x^{11})(1+x^{15})(1+x^{19})(1+x^{23})(1+x^{27})(1+x^{35})$$

$$P_{E_8}(x) = (1+x^3)(1+x^{15})(1+x^{23})(1+x^{27})(1+x^{35})(1+x^{39})(1+x^{47})(1+x^{59})$$

Relationship with Dimensions of Spaces of Differential Forms

In geometric situations when X is a closed manifold, the importance of the Betti numbers may arise from a different direction, namely that they predict the dimensions of vector spaces of closed differential forms *modulo* exact differential forms.

The connection with the definition given above is via three basic results, de Rham's theorem and Poincaré duality (when those apply), and the universal coefficient theorem of homology theory.

There is an alternate reading, namely that the Betti numbers give the dimensions of spaces of harmonic forms. This requires also the use of some of the results of Hodge theory, about the Hodge Laplacian.

In this setting, Morse theory gives a set of inequalities for alternating sums of Betti numbers in terms of a corresponding alternating sum of the number of critical points *Ni* of a Morse function of a given index:

$$b_i(X) - b_{i-1}(X) + \cdots \le N_i - N_{i-1} + \cdots.$$

Witten gave an explanation of these inequalities by using the Morse function to modify the exterior derivative in the de Rham complex.

Riemann–Hurwitz Formula

In mathematics, the Riemann–Hurwitz formula, named after Bernhard Riemann and Adolf Hurwitz, describes the relationship of the Euler characteristics of two surfaces when one is a *ramified covering* of the other.

It therefore connects ramification with algebraic topology, in this case. It is a prototype result for many others, and is often applied in the theory of Riemann surfaces (which is its origin) and algebraic curves.

Statement

For an orientable surface S the Euler characteristic $\chi(S)$ is

$$2 - 2g$$

where g is the genus (the *number of handles*), since the Betti numbers are 1, $2g$, 1, 0, 0, ... In the case of an (*unramified*) covering map of surfaces

$$\pi : S' \to S$$

that is surjective and of degree N, we should have the formula

$$\chi(S') = N \cdot \chi(S)$$

That is because each simplex of S should be covered by exactly N in $S2$ —at least if we use a fine enough triangulation of S, as we are entitled to do since the Euler characteristic is a topological invariant. What the Riemann–Hurwitz formula does is to add in a correction to allow for ramification (*sheets coming together*).

Now assume that S and $S2$ are Riemann surfaces, and that the map π is complex analytic. The map π is said to be *ramified* at a point P in $S2$ if there exist analytic coordinates near P and $\pi(P)$ such that π takes the form $\pi(z) = z^n$, and $n > 1$. An equivalent way of thinking about this is that there exists a small neighbourhood U of P such that $\pi(P)$ has exactly one preimage in U, but the image of any other point in U has exactly n preimages in U.

The number n is called the *ramification index at P* and also denoted by e_P. In calculating the Euler characteristic of $S2$ we notice the loss of e_P - 1 copies of P above $\pi(P)$ (that is, in the inverse image of $\pi(P)$).

Now let us choose triangulations of S and $S2$ with vertices at the branch and ramification points, respectively, and use these to compute the Euler characteristics.

Then $S2$ will have the same number of d-dimensional faces for d different from zero, but fewer than expected vertices. Therefore we find a "corrected" formula

$$\chi(S') = N{\cdot}\chi(S) - \sum_{P \in S'} (e_P - 1)$$

(almost all P have $e_P = 1$, so this is quite safe). This formula is known as the *Riemann–Hurwitz formula* and also as Hurwitz's theorem.

Examples

The Weierstrass -function, considered as a meromorphic function with values in the Riemann sphere, yields a map from an elliptic curve (genus 1) to the projective line (genus 0). It is a double cover ($N = 2$), with ramification at four points only, at which $e = 2$. The Riemann–Hurwitz formula then reads

$$0 = 2 \cdot 2 - \Sigma\ 1$$

with the summation taken over four values of P.

The formula may also be used to calculate the genus of hyperelliptic curves.

As another example, the Riemann sphere maps to itself by the function z^n, which has ramification index n at 0, for any integer $n > 1$. There can only be other ramification at the point at infinity. In order to balance the equation

$$2 = n \cdot 2 - (n - 1) - (\geq - 1)$$

we must have ramification index n at infinity, also.

Consequences

Several results in algebraic topology and complex analysis follow.

Firstly, there are no ramified covering maps from a curve of lower genus to a curve of higher genus – and thus, since non-constant meromorphic maps of curves are ramified covering spaces, there are no non-constant meromorphic maps from a curve of lower genus to a curve of higher genus.

As another example, it shows immediately that a curve of genus 0 has no cover with $N > 1$ that is unramified everywhere: because that would give rise to an Euler characteristic > 2.

Generalizations

For a correspondence of curves, there is a more general formula, Zeuthen's theorem, which gives the ramification correction to the first approximation that the Euler characteristics are in the inverse ratio to the degrees of the correspondence.

Mayer–Vietoris Sequence

In mathematics, particularly algebraic topology and homology theory, the Mayer–Vietoris sequence is an algebraic tool to help compute algebraic invariants of topological spaces, known as their homology and cohomology groups. The result is due to two Austrian mathematicians, Walther Mayer and Leopold Vietoris. The method consists of splitting a space into pieces, called subspaces, for which the homology or cohomology groups may be easier to compute.

The sequence relates the (co)homology groups of the space to the (co)homology groups of the subspaces. It is a natural long exact sequence, whose entries are the (co)homology groups of the whole space, the direct sum of the (co)homology groups of the subspaces, and the (co)homology groups of the intersection of the subspaces.

The Mayer–Vietoris sequence holds for a variety of cohomology and homology theories, including singular homology and singular cohomology. In general, the sequence holds for those theories satisfying the Eilenberg-Steenrod axioms, and it has variations for both reduced and relative (co)homology.

Because the (co)homology of most spaces cannot be computed directly from their definitions, one uses tools such as the Mayer–Vietoris sequence in the hope of obtaining partial information.

Many spaces encountered in topology are constructed by piecing together very simple patches. Carefully choosing the two covering subspaces so that, together with their intersection, they have simpler (co)homology than that of the whole space may allow a complete deduction of the (co)homology of the space. In that respect, the Mayer–Vietoris sequence is analogous to the Seifert–van Kampen theorem for the fundamental group, and a precise relation exists for homology of dimension one.

Background, Motivation, and History

Like the fundamental group or the higher homotopy groups of a space, homology groups are important topological invariants. Although some (co)homology theories are computable using tools of linear algebra, many other important (co)homology theories, especially singular (co)homology, are not computable directly from their definition for nontrivial spaces. For singular (co)homology, the singular (co)chains and (co)cycles groups are often too big to handle directly. More subtle and indirect approaches become necessary. The Mayer–Vietoris sequence is such an approach, giving partial information about the (co)homology groups of any space by relating it to the (co)homology groups of two of its subspaces and their intersection.

The most natural and convenient way to express the relation involves the algebraic concept of exact sequences: sequences of objects (in this case groups) and morphisms (in this case group homomorphisms) between them such that the image of one morphism equals the kernel of the next. In general, this does not allow (co)homology groups of a space to be completely computed. However, because many important spaces encountered in topology are topological manifolds, simplicial complexes, or CW complexes, which are constructed by piecing together very simple patches, a theorem such as that of Mayer and Vietoris is potentially of broad and deep applicability.

Mayer was introduced to topology by his colleague Vietoris when attending his lectures in 1926 and 1927 at a local university in Vienna. He was told about the conjectured result and a way to its solution, and solved the question for the Betti numbers in 1929. He applied his results to the torus considered as the union of two cylinders. Vietoris later proved the full result for the homology groups in 1930 but did not express it as an exact sequence.

The concept of an exact sequence only appeared in print in the 1952 book *Foundations of Algebraic Topology* by Samuel Eilenberg and Norman Steenrod where the results of Mayer and Vietoris were expressed in the modern form.

Basic Versions for Singular Homology

Let X be a topological space and A, B be two subspaces whose interiors cover X. (The interiors of A and B need not be disjoint.) The Mayer–Vietoris sequence in singular homology for the triad (X, A, B) is a long exact sequence relating the singular homology groups (with coefficient group the integers Z) of the spaces X, A, B, and the intersection *A*)”*B*. There is an unreduced and a reduced version.

Boundary Map

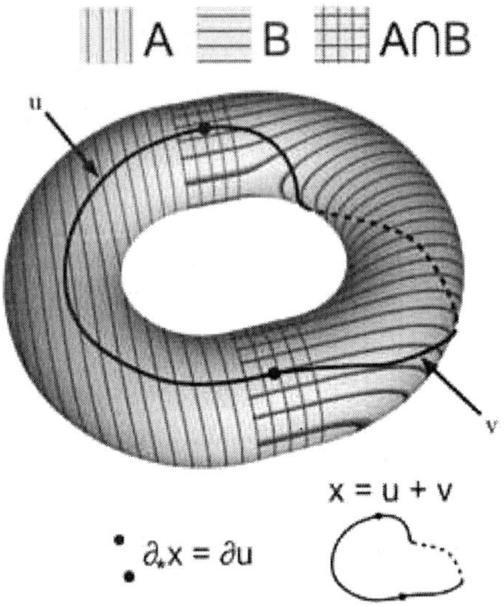

Figure: *Illustration of the boundary map ∂_* on the torus where the 1-cycle* x = u + v *is the sum of two 1-chains whose boundary lies in the intersection of* A *and* B.

The boundary maps "$_*$ lowering the dimension may be made explicit as follows. An element in $H_n(X)$ is the homology class of an n-cycle x which, by barycentric subdivision for example, can be written as the sum of two n-chains u and v whose images lie wholly in A and B, respectively. Thus $\partial x = \partial(u + v) = 0$ so that $\partial u = \partial v$.

This implies that the images of both these boundary (n - 1)-cycles are contained in the intersection $A \cap B$. Then $\partial_*([x])$ is the class of $-u$ in $H_{n-1}(A \cap B)$. Choosing a different representative $x2$ does not affect $-u$ since $-x2 = \partial x = 0$; nor does choosing another decomposition $x = u2 + v2$ since then $\partial u + \partial v + \partial ut - \partial vt = 0$ which implies $\partial u = \partial u2$ and $\partial v = \partial v2$. Notice that the maps in the Mayer–Vietoris sequence depend on choosing an order for A and B. In particular, the boundary map changes sign if A and B are swapped.

Reduced Version

For reduced homology there is also a Mayer–Vietoris sequence, under the assumption that A and B have non-empty intersection. The sequence is identical for positive dimensions and ends as:

$$\cdots \to \tilde{H}_0(A \cap B) \xrightarrow{(i_*, j_*)} \tilde{H}_0(A) \oplus \tilde{H}_0(B) \xrightarrow{k_* - l_*} \tilde{H}_0(X) \to 0.$$

Analogy with the Seifert–Van Kampen Theorem

There is an analogy between the Mayer–Vietoris sequence (especially for homology groups of dimension 1) and the Seifert–van Kampen theorem. Whenever A)"B is path-connected the reduced Mayer–Vietoris sequence yields the isomorphism

$$H_1(X) \cong (H_1(A) \oplus II_1(B)) / \mathrm{Ker}(k_* - l_*)$$

where, by exactness,

$$\mathrm{Ker}(k_* - l_*) \cong \mathrm{Im}(i_*, j_*).$$

This is precisely the abelianized statement of the Seifert–van Kampen theorem. Compare with the fact that $H_1(X)$ is the abelianization of the fundamental group $\pi_1(X)$ when X is path-connected.

Universal Coefficient Theorem

In mathematics, the universal coefficient theorem in algebraic topology establishes the relationship in homology theory between the *integral homology* of a topological space X, and its *homology with coefficients* in any abelian group A. It states that the integral homology groups

$H_i(X, \mathrm{Z})$

completely determine the groups

$H_i(X, A)$

Here H_i might be the simplicial homology or more general singular homology theory: the result itself is a pure piece of homological algebra about chain complexes of free abelian groups. The form of the result is that other coefficients A may be used, at the cost of using a Tor functor.

For example it is common to take A to be Z/2Z, so that coefficients are modulo 2. This becomes straightforward in the absence of 2-torsion in the homology. Quite generally, the result indicates the relationship that holds between the Betti numbers b_i of X and the Betti numbers $b_{i,F}$ with coefficients in a field F. These can differ, but only when the characteristic of F is a prime number p for which there is some p-torsion in the homology.

Statement

Consider the tensor product of modules $H_i(X, \mathrm{Z}) \otimes A$. The theorem states that there is an injective group homomorphism é from this group to $H_i(X, A)$, which has cokernel Tor($H_{i-1}(X, \mathrm{Z}), A$). In other words, there is a natural short exact sequence

$$0 \to H_i(X,\mathbf{Z}) \otimes A \to H_i(X,A) \to \mathrm{Tor}(H_{i-1}(X,\mathbf{Z}),A) \to 0.$$

Furthermore, this is a split sequence (but the splitting is *not* natural). The Tor group on the right can be thought of as the obstruction to é being an isomorphism.

Universal Coefficient Theorem for Cohomology

There is also a universal coefficient theorem for cohomology involving the Ext functor, stating that there is a natural short exact sequence

$$0 \to \mathrm{Ext}(H_{i-1}(X,\mathbf{Z}),A) \to H^i(X,A) \to \mathrm{Hom}(H_i(X,\mathbf{Z}),A) \to 0.$$

As in the homological case, the sequence splits, though not naturally.

Example: Mod 2 Cohomology of the Real Projective Space

Let $X = P^n(R)$, the real projective space. We compute the singular cohomology of X with coefficients in $R := Z_2$.

Knowing that the integer homology is given by:

$$H_i(X;\mathbf{Z}) = \begin{cases} \mathbf{Z} & i = 0 \text{ or } i = n \text{ odd}, \\ \mathbf{Z}/2\mathbf{Z} & 0 < i < n, i \text{ odd}, \\ 0 & \text{else.} \end{cases}$$

We have Ext(R, R) = R, Ext(Z, R)= 0, so that the above exact sequences yield

$$\forall i = 0 \ldots n, H^i(X;R) = R.$$

In fact the total cohomology ring structure is

$$H^*(X;R) = R[w]/\langle w^{n+1}\rangle.$$

Corollaries

A special case of the theorem is computing integral cohomology. For a finite CW complex X, $H_i(X;\mathbf{Z})$ is finitely generated, and so we have the following decomposition.

$$H_i(X;\mathbf{Z}) \cong \mathbf{Z}^{\beta_i(X)} \oplus T_i.$$

Where $\beta_i(X)$ are the betti numbers of X. One may check that

$$\text{Hom}(H_i(X),\mathbf{Z}) \cong \text{Hom}(\mathbf{Z}^{\beta_i(X)}) \oplus \text{Hom}(T_i,\mathbf{Z}) \cong \mathbf{Z}^{\beta_i(X)}$$

and

$$\text{Ext}(H_i(X),\mathbf{Z}) \cong \text{Ext}(\mathbf{Z}^{\beta_i(X)}) \oplus \text{Ext}(T_i,\mathbf{Z}) \cong T_i.$$

This gives the following statement for integral cohomology:

$$H^i(X;\mathbf{Z}) \cong \mathbf{Z}^{\beta_i(X)} \oplus T_{i-1}.$$

For X a compact, closed, and connected n-manifold, this corollary coupled with Poincaré duality gives that $\beta_i(X) = \beta_{n-i}(X)$.

Cup Product

In mathematics, specifically in algebraic topology, the cup product is a method of adjoining two cocycles of degree p and

q to form a composite cocycle of degree $p + q$. This defines an associative (and distributive) graded commutative product operation in cohomology, turning the cohomology of a space X into a graded ring, $H^{\cdot}(X)$, called the cohomology ring.

The cup product was introduced in work of J. W. Alexander, Eduard Èech and Hassler Whitney from 1935–1938, and, in full generality, by Samuel Eilenberg in 1944.

Definition

In singular cohomology, the cup product is a construction giving a product on the graded cohomology ring $H^{\cdot}(X)$ of a topological space X. The construction starts with a product of cochains: if c^p is a p-cochain and d^q is a q-cochain, then

$$(c^p \smile d^q)(\sigma) = c^p(\sigma \circ \iota_{0,1,\dots p}) \cdot d^q(\sigma \circ \iota_{p,p+1,\dots,p+q})$$

where σ is a $(p + q)$ -singular simplex and $\iota_S, S \subset \{0,1,\dots,p+q\}$ is the canonical embedding of the simplex spanned by S into the $(p+q)$-standard simplex.

Informally, $\sigma \circ \iota_{0,1,\dots,p}$ is the p-th front face and $\sigma \circ \iota_{p,p+1,\dots,p+q}$ is the q-th back face of σ, respectively. The coboundary of the cup product of cocycles c^p and d^q is given by

$$\delta(c^p \smile d^q) = \delta c^p \smile d^q + (-1)^p (c^p \smile \delta d^q).$$

The cup product of two cocycles is again a cocycle, and the product of a coboundary with a cocycle (in either order) is a coboundary. Thus, the cup product operation passes to cohomology, defining a bilinear operation

$$H^p(X) \times H^q(X) \to H^{p+q}(X).$$

Properties

The cup product operation in cohomology satisfies the identity

$$\alpha^p \smile \beta^q = (-1)^{pq}(\beta^q \smile \alpha^p)$$

so that the corresponding multiplication is graded-commutative.

The cup product is functorial, in the following sense: if

$$f : X \to Y$$

is a continuous function, and

$$f^*: H^*(Y) \to H^*(X)$$

is the induced homomorphism in cohomology, then

$$f^*(\alpha \smile \beta) = f^*(\alpha) \smile f^*(\beta),$$

for all classes α, β in $H^*(Y)$. In other words, f^* is a (graded) ring homomorphism.

Interpretation

It is possible to view the cup product $\smile: H^p(X) \times H^q(X) \to H^{p+q}(X)$ as induced from the following composition:

$$C^{\cdot}(X) \times C^{\cdot}(X) \to C^{\cdot}(X \times X) \xrightarrow{\Delta^*} C^{\cdot}(X)$$

in terms of the chain complexes of X and $X \times X$, where the first map is the Künneth map and the second is the map induced by the diagonal $\Delta: X \to X \times X$.

This composition passes to the quotient to give a well-defined map in terms of cohomology, this is the cup product. This approach explains the existence of a cup product for cohomology but not for homology: $\Delta: X \to X \times X$ induces a map $\Delta^*: H^{\cdot}(X \times X) \to H^{\cdot}(X)$ but would also induce a map $\Delta_*: H_{\cdot}(X) \to H_{\cdot}(X \times X)$, which goes the wrong way round to allow us to define a product.

This is however of use in defining the cap product. Bilinearity follows from this presentation of cup product, i.e. $(u_1 + u_2) \smile v = u_1 \smile v + u_2 \smile v$ and $u \smile (v_1 + v_2) = u \smile v_1 + u \smile v_2$.

Examples

Cup products may be used to distinguish manifolds from wedges of spaces with identical cohomology groups. First consider the Torus, T. One may check that if $E = S^1 \vee S^1$, then $X := S^2 \vee E$ has identical cohomology groups.

However, the multiplication of the cup product distinguishes the associated cohomology rings. In the case of X the multiplication of the cochains associated to the copies of S^1 is degenerate, whereas in T multiplication in the first cohomology group can be used to decompose the torus as a 2-cell diagram, thus having product equal to Z (more generally M where this is the base module).

Other Definitions

Cup Product and Differential Forms

In de Rham cohomology, the cup product of differential forms is also known as the wedge product, and in this sense is a special case of Grassmann's exterior product.

Cup Product and Geometric Intersections

When two submanifolds of a smooth manifold intersect transversely, their intersection is again a submanifold. By taking the fundamental homology class of these manifolds, this yields a bilinear product on homology.

This product is dual to the cup product, i.e. the homology class of the intersection of two submanifolds is the Poincaré dual of the cup product of their Poincaré duals.

Similarly, the linking number can be defined in terms of intersections, shifting dimensions by 1, or alternatively in terms of a non-vanishing cup product on the complement of a link.

Massey Products

The cup product is a binary (2-ary) operation; one can define a ternary (3-ary) and higher order operation called the Massey product, which generalizes the cup product. This is a higher order cohomology operation, which is only partly defined (only defined for some triples).

Cohomology Ring

In mathematics, specifically algebraic topology, the cohomology ring of a topological space X is a ring formed from the cohomology groups of X together with the cup product serving as the ring multiplication.

Here 'cohomology' is usually understood as singular cohomology, but the ring structure is also present in other theories such as de Rham cohomology. It is also functorial: for a continuous mapping of spaces one obtains a ring homomorphism on cohomology rings, which is contravariant.

Specifically, given a sequence of cohomology groups $H^k(X;R)$ on X with coefficients in a commutative ring R (typically R is Z_n, Z, Q, R, or C) one can define the cup product, which takes the form

$$H^k(X;R) \times H^\ell(X;R) \to H^{k+\ell}(X;R).$$

The cup product gives a multiplication on the direct sum of the cohomology groups

$$H^\cdot(X;R) = \bigoplus_{k \in \mathbb{N}} H^k(X;R).$$

This multiplication turns $H^\cdot(X;R)$ into a ring. If fact, it is naturally an N-graded ring with the nonnegative integer k serving as the degree. The cup product respects this grading.

The cohomology ring is graded-commutative in the sense that the cup product commutes up to a sign determined by the grading. Specifically, for pure elements of degree k and –!; we have

$$(\alpha^k \smile \beta^\ell) = (-1)^{k\ell}(\beta^\ell \smile \alpha^k).$$

A numerical invariant derived from the cohomology ring is the cup-length, which means the maximum number of graded elements of degree e– 1 that when multiplied give a non-zero result. For example a complex projective space has cup-length equal to its complex dimension.

Chapter 6

Topological Ring

In mathematics, a topological ring is a ring R which is also a topological space such that both the addition and the multiplication are continuous as maps

$$R \times R \to R,$$

where $R \times R$ carries the product topology.

General Comments

The group of units of R may not be a topological group using the subspace topology, as inversion on the unit group need not be continuous with the subspace topology. (An example of this situation is the adele ring of a global field. Its unit group, called the idele group, is not a topological group in the subspace topology.) Embedding the unit group of R into the product $R \times R$ as (x,x^{-1}) does make the unit group a topological group. (If inversion on the unit group is continuous in the subspace topology of R then the topology on the unit group viewed in R or in $R \times R$ as above are the same.)

If one does not require a ring to have a unit, then one has to add the requirement of continuity of the additive inverse, or equivalently, to define the topological ring as a ring which is a topological group (for +) in which multiplication is continuous, too.

Examples

Topological rings occur in mathematical analysis, for examples as rings of continuous real-valued functions on some topological space (where the topology is given by pointwise convergence), or as rings of continuous linear operators on some normed vector space; all Banach algebras are topological rings. The rational, real, complex and p-adic numbers are also topological rings (even topological fields) with their standard topologies. In the plane, split-complex numbers and dual numbers form alternative topological rings.

Completion (Ring Theory)

In abstract algebra, a completion is any of several related functors on rings and modules that result in complete topological rings and modules. Completion is similar to localization, and together they are among the most basic tools in analysing commutative rings.

Complete commutative rings have simpler structure than the general ones and Hensel's lemma applies to them. Geometrically, a completion of a commutative ring R concentrates on a formal neighbourhood of a point or a Zariski closed subvariety of its spectrum Spec R.

General Construction

Suppose that E is an abelian group with a descending filtration

$$E = F^0E \supset F^1E \supset F^2E \supset \cdots$$

of subgroups, one defines the completion (with respect to the filtration) as the inverse limit:

$$\hat{E} = \varprojlim(E / F^n E).$$

This is again an abelian group. Usually E is an *additive* abelian group. If E has additional algebraic structure compatible with the filtration, for instance E is a filtered ring, a filtered module, or a filtered vector space, then its completion is again an object with the same structure that is complete in the topology determined by the filtration. This construction may be applied both to commutative and noncommutative rings. As may be expected, this produces a complete topological ring.

Krull Topology

In commutative algebra, the filtration on a commutative ring R by the powers of a proper ideal I determines the Krull topology (after Wolfgang Krull) or I-adic topology on R. The case of a *maximal* ideal $I = \mathfrak{m}$ is especially important.

The basis of open neighbourhoods of 0 in R is given by the powers I^n, which are *nested* and form a descending filtration on R:

$$F^0R = R \supset I \supset I^2 \supset \cdots, \quad F^nR = I^n.$$

The completion is the inverse limit of the factor rings,

$$\hat{R}_I = \varprojlim(R / I^n)$$

(pronounced "R I hat"). The kernel of the canonical map π from the ring to its completion is the intersection of the powers of I. Thus π is injective if and only if this intersection reduces to the zero element of the ring; by the Krull intersection theorem, this is the case for any commutative Noetherian ring which is either an integral domain or a local ring.

There is a related topology on R-modules, also called Krull or I-adic topology. A basis of open neighbourhoods of a module M is given by the sets of the form

$$x + I^nM \quad \text{for } x \in M.$$

The completion of an R-module M the inverse limit of the quotients

$$\hat{M}_I = \varprojlim(M / I^nM).$$

This procedure converts any module over R into a complete topological module over $\hat{R}_I$.

Examples

1. The ring of p-adic integers Z_p is obtained by completing the ring Z of integers at the ideal (p).
2. Let $R = K[x_1,\ldots,x_n]$ be the polynomial ring in n variables over a field K and $\mathfrak{m} = (x_1,\ldots,x_n)$ be the maximal ideal generated by the variables. Then the completion R_m is the ring $K[[x_1,\ldots,x_n$ of formal power series in n variables over K.
3. Let R be the ring of holomorphic functions on a complex manifold and let I be the maximal ideal of functions vanishing at some point p. Then the completion of R at the ideal I is the ring of power series over C that are convergent in a neighbourhood of p.

Topological Fields

Some of the most important examples are also fields F. To have a topological field we should also specify that inversion is continuous, when restricted to $F\setminus\{0\}$.

Pointed Space

In mathematics, a pointed space is a topological space X with a distinguished basepoint x_0 in X. Maps of pointed spaces (based maps) are continuous maps preserving basepoints, i.e. a continuous map $f : X \to Y$ such that $f(x_0) = y_0$. This is usually denoted

$$f : (X, x_0) \to (Y, y_0).$$

Pointed spaces are important in algebraic topology, particularly in homotopy theory, where many constructions, such as the fundamental group, depend on a choice of basepoint. The pointed set concept is less important; it is anyway the case of a pointed discrete space.

Category of Pointed Spaces

The class of all pointed spaces forms a category Top. with basepoint preserving continuous maps as morphisms. Another way to think about this category is as the comma category, ({•} –! Top) where {•} is any one point space and Top is the category of topological spaces.

(This is also called a coslice category denoted {•}/Top.) Objects in this category are continuous maps $\{\bullet\} \to X$. Such morphisms can be thought of as picking out a basepoint in X. Morphisms in ({•} –! Top) are morphisms in Top for which the following diagram commutes:

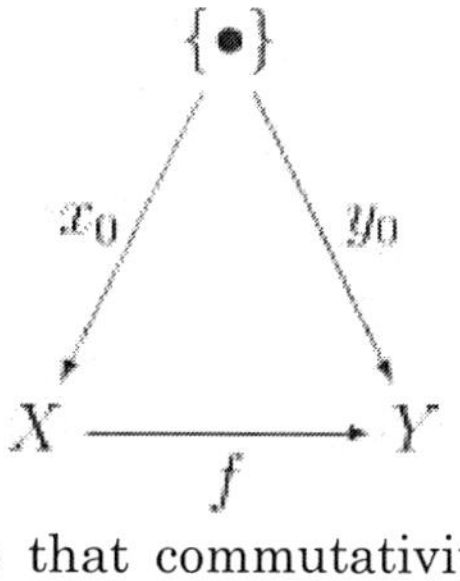

It is easy to see that commutativity of the diagram is equivalent to the condition that f preserves basepoints.

As a pointed space {•} is a zero object in Top$_{.}$ while it is only a terminal object in Top.

There is a forgetful functor Top$_{.}$ → Top which "forgets" which point is the basepoint.

This functor has a left adjoint which assigns to each topological space X the disjoint union of X and a one point space {•} whose single element is taken to be the basepoint.

Adjunction Space

In mathematics, an adjunction space (or attaching space) is a common construction in topology where one topological space is attached or "glued" onto another. Specifically, let X and Y be a topological spaces with A a subspace of Y. Let $f : A \to X$ be a continuous map (called the attaching map). One forms the adjunction space X *"$_f$ Y by taking the disjoint union of X and Y and identifying x with $f(x)$ for all x in A. Schematically,

$$X \cup_f Y = (X \coprod Y) / \{f(A) \sim A\}.$$

Sometimes, the adjunction is written as $X +_f Y$. Intuitively, we think of Y as being glued onto X via the map f. As a set, $X \cup_f Y$ consists of the disjoint union of X and (Y – A). The topology, however, is specified by the quotient construction. In the case where A is a closed subspace of Y one can show that the map $X \to X \cup_f$ is a closed embedding and $(Y - A) \to X \cup_f Y$ is an open embedding.

Semi-Locally Simply Connected

In mathematics, specifically algebraic topology, the phrase semi-locally simply connected refers to a certain local connectedness condition that arises in the theory of covering spaces. Roughly speaking, a topological space X is semi-locally simply connected if there is a lower bound on the sizes of the "holes" in X. This condition is necessary for most of the theory of covering spaces, including the existence of a universal cover and the Galois correspondence between covering spaces and subgroups of the fundamental group. Most "nice" spaces such as manifolds and CW complexes are semi-locally simply connected, and topological spaces that do not satisfy this condition are considered somewhat pathological. The standard example of a non-semi-locally simply connected space is the Hawaiian earring.

Definition

A space X is called semi-locally simply connected if every point in X has a neighbourhood U with the property that every loop in U can be contracted to a single point within X (i.e. every loop is nullhomotopic). Note that the neighbourhood U need not be simply connected: though every loop in U must be contractible within X, the contraction is not required to take place inside of U. For this reason, a space can be semi-locally simply connected without being locally simply connected.

Equivalent to this definition, a space X is semi-locally simply connected if every point in X has a neighbourhood U for which the homomorphism from the fundamental group of U to the fundamental group of X, induced by the inclusion map of U into X, is trivial. Most of the main theorems about covering spaces, including the existence of a universal cover and the Galois correspondence, require a space to be path-connected, locally path-connected, and semi-locally simply connected. In particular, this condition is necessary for a space to have a simply connected covering space.

Examples

Any "nice" space such as a manifold or CW complex is semi-locally simply connected. In some sense, only a pathological space can fail to satisfy this condition. A simple example of a space that is not semi-locally simply connected is the Hawaiian earring: the union of tho circles in the Euclidean plane with centres $(1/n, 0)$ and radii $1/n$, for n a natural number. Give this space the subspace topology. Then all neighbourhoods of the origin contain circles that are not nullhomotopic.

The Hawaiian earring can also be used to construct a semi-locally simply connected space that is not locally simply connected. In particular, the cone on the Hawaiian earring is contractible and therefore semi-locally simply connected, but it is clearly not locally simply connected.

Another example of a non-semi-locally simply connected space is the complement of $Q \times Q$ in the Euclidean plane R^2, where Q denotes the set of rational numbers. In fact, the fundamental group of this space is uncountable (Hatcher p. 54).

Topology of Fundamental Group

In terms of the natural topology on the fundamental group, a space is semi-locally simply connected if and only if its topological fundamental group is discrete.

Polytope

In elementary geometry, a polytope is a geometric object with flat sides, which exists in any general number of dimensions. A polygon is a polytope in two dimensions, a polyhedron in three dimensions, and so on in higher dimensions (such as a polychoron in four dimensions). Some theories further generalize the idea to include such things as unbounded polytopes (apeirotopes and tessellations), and abstract polytopes.

When referring to an n-dimensional generalization, the term n-polytope is used. For example, a polygon is a 2-polytope, a polyhedron is a 3-polytope, and a polychoron is a 4-polytope.

The term was coined by the mathematician Hoppe, writing in German, and was later introduced to English by Alicia Boole Stott, the daughter of logician George Boole.

Different Approaches to Definition

The term *polytope* is a broad term that covers a wide class of objects, and different definitions are attested in mathematical literature. Many of these definitions are not equivalent, resulting in different sets of objects being called *polytopes*. They represent different approaches of generalizing the convex polytopes to include other objects with similar properties and aesthetic beauty. The original approach broadly followed by Schläfli, Gossett and others begins with the 0-dimensional point as a 0-polytope (vertex). A 1-dimensional 1-polytope (edge) is constructed by bounding a line segment with two 0-polytopes. Then 2-polytopes (polygons) are defined as plane objects whose bounding facets (edges) are 1-polytopes, 3-polytopes (polyhedra) are defined as solids whose facets (faces) are 2-polytopes, and so forth.

A polytope may also be regarded as a tessellation of some given manifold. Convex polytopes are equivalent to tilings of the sphere, while others may be tilings of other elliptic, flat or toroidal surfaces. Under this definition, plane tilings and space tilings (honeycombs) are considered to be polytopes, and are sometimes classed as apeirotopes because they have infinitely many cells; tilings of hyperbolic spaces are also included under this definition.

An alternative approach defines a polytope as a set of points that admits a simplicial decomposition. In this definition, a polytope is the union of finitely many simplices, with the additional property that, for any two simplices that have a nonempty intersection, their intersection is a vertex, edge, or higher dimensional face of the two. However this definition does not allow star polytopes with interior structures, and so is restricted to certain areas of mathematics.

The theory of abstract polytopes attempts to detach polytopes from the space containing them, considering their purely combinatorial properties. This allows the definition of the term to be extended to include objects for which it is difficult to define clearly a *natural underlying space*, such as the 11-cell.

Elements

The elements of a polytope are its vertices, edges, faces, cells and so on. The terminology for these is not entirely consistent across different authors. To give just a few examples: Some authors use *face* to refer to an (n–1)-dimensional element while others use *face* to denote a 2-face specifically, and others use j-face or k-face to indicate an element of j or k dimensions. Some sources use *edge* to refer to a ridge, while H. S. M. Coxeter uses *cell* to denote an (n–1)-dimensional element.

An n-dimensional polytope is bounded by a number of (n–1)-dimensional *facets*. These facets are themselves polytopes, whose facets are (n–2)-dimensional *ridges* of the original polytope. Every ridge arises as the intersection of two facets (but the intersection of two facets need not be a ridge). Ridges are once again polytopes whose facets give rise to (n–3)-dimensional boundaries of the original polytope, and so on. These bounding sub-polytopes may be referred to as faces, or specifically j-dimensional faces or j-faces. A 0-dimensional face is called a *vertex*, and consists of a single point. A 1-dimensional face is called an *edge*, and consists of a line segment. A 2-

dimensional face consists of a polygon, and a 3-dimensional face, sometimes called a *cell*, consists of a polyhedron.

Special Classes of Polytope

Regular Polytopes

A polytope may be *regular*. The regular polytopes are a class of highly symmetrical and aesthetically pleasing polytopes, including the Platonic solids, which have been studied extensively since ancient times.

Convex Polytopes

A polytope may be *convex*. The convex polytopes are the simplest kind of polytopes, and form the basis for different generalizations of the concept of polytopes. A convex polytope is sometimes defined as the intersection of a set of half-spaces. This definition allows a polytope to be neither bounded nor finite. Polytopes are defined in this way, e.g., in linear programming. A polytope is *bounded* if there is a ball of finite radius that contains it. A polytope is said to be *pointed* if it contains at least one vertex. Every bounded nonempty polytope is pointed. An example of a non-pointed polytope is the set $\{(x, y) \in \mathbb{R}^2 \mid x \geq 0\}$. A polytope is *finite* if it is defined in terms of a finite number of objects, e.g., as an intersection of a finite number of half-planes.

Star Polytopes

A non-convex polytope may be self-intersecting; this class of polytopes include the star polytopes.

Abstract Polytopes

An abstract polytope is a partially ordered set of elements or members, which obeys certain rules. It is a purely algebraic structure, and the theory was developed in order to avoid some of the issues which make it difficult to reconcile the various geometric classes within a consistent mathematical framework. A geometric polytope is said to be a realisation of some associated abstract polytope.

Self-Dual Polytopes

In 2 dimensions, all regular polygons (regular 2-polytopes) are self-dual.

In 3 dimensions, the tetrahedron is self-dual, as well as canonical polygonal pyramids and elongated pyramids.

In higher dimensions, every regular n-simplex, with Schlafli symbol $\{3^n\}$, is self-dual.

In addition, the 24-cell in 4 dimensions, with Schlafli symbol {3,4,3}, is self-dual.

History

The concept of a polytope originally began with polygons and polyhedra, both of which have been known since ancient times:

It was not until the 19th century that higher dimensions were discovered and geometres learned to construct analogues of polygons and polyhedra in them. The first hint of higher dimensions seems to have come in 1827, with Möbius' discovery that two mirror-image solids can be superimposed by rotating one of them through a fourth dimension. By the 1850s, a handful of other mathematicians such as Cayley and Grassman had considered higher dimensions. Ludwig Schläfli was the first of these to consider analogues of polygons and polyhedra in such higher spaces. In 1852 he described the six convex regular 4-polytopes, but his work was not published until 1901, six years after his death. By 1854, Bernhard Riemann's *Habilitationsschrift* had firmly established the geometry of higher dimensions, and thus the concept of n-dimensional polytopes was made acceptable. Schläfli's polytopes were rediscovered many times in the following decades, even during his lifetime.

In 1882 Hoppe, writing in German, coined the word *polytop* to refer to this more general concept of polygons and polyhedra. In due course, Alicia Boole Stott introduced *polytope* into the English language.

In 1895, Thorold Gosset not only rediscovered Schläfli's regular polytopes, but also investigated the ideas of semiregular polytopes and space-filling tessellations in higher dimensions. Polytopes were also studied in non-Euclidean spaces such as hyperbolic space.

During the early part of the 20th century, higher-dimensional spaces became fashionable, and together with the idea of higher polytopes, inspired artists such as Picasso to create the movement known as cubism.

An important milestone was reached in 1948 with H. S. M. Coxeter's book *Regular Polytopes*, summarising work to date and adding findings of his own. Branko Grünbaum published his influential work on *Convex Polytopes* in 1967.

More recently, the concept of a polytope has been further generalized. In 1952 Shephard developed the idea of complex polytopes in complex space, where each real dimension has an imaginary one associated with it. Coxeter went on to publish his book, *Regular Complex Polytopes*, in 1974.

Complex polytopes do not have closed surfaces in the usual way, and are better understood as configurations. This kind of conceptual issue led to the more general idea of incidence complexes and the study of abstract combinatorial properties relating vertices, edges, faces and so on.

This in turn led to the theory of abstract polytopes as partially ordered sets, or posets, of such elements. McMullen and Schulte published their book *Abstract Regular Polytopes* in 2002.

Enumerating the uniform polytopes, convex and nonconvex, in four or more dimensions remains an outstanding problem.

In modern times, polytopes and related concepts have found many important applications in fields as diverse as computer graphics, optimization, search engines, cosmology and numerous other fields.

Uses

In the study of optimization, linear programming studies the maxima and minima of linear functions constricted to the boundary of an n-dimensional polytope.

In linear programming, polytopes occur in the use of Generalized barycentric coordinates and Slack variables.

Disjoint Union (Topology)

In general topology and related areas of mathematics, the disjoint union (also called the direct sum, free union, free sum, topological sum, or coproduct) of a family of topological spaces is a space formed by equipping the disjoint union of the underlying sets with a natural topology called the disjoint union topology.

Roughly speaking, two or more spaces may be considered together, each looking as it would alone. The name *coproduct* originates from the fact that the disjoint union is the categorical dual of the product space construction.

Definition

Let $\{X_i : i \in I\}$ be a family of topological spaces indexed by I. Let

$$X = \coprod_i X_i$$

be the disjoint union of the underlying sets. For each i in I, let

$$\varphi_i : X_i \to X$$

be the canonical injection (defined by $\varphi_i(x) = (x, i)$). The disjoint union topology on X is defined as the finest topology on X for which the canonical injections are continuous (i.e. the final topology for the family of functions $\{\varphi_i\}$).

Explicitly, the disjoint union topology can be described as follows. A subset U of X is open in X if and only if its preimage $\varphi_i^{-1}(U$ is open in X_i for each $i \in I$.

Yet another formulation is that a subset V of X is open relative to X iff its intersection with X_i is open relative to X_i for each i.

Examples

If each X_i is homeomorphic to a fixed space A, then the disjoint union X will be homeomorphic to $A \times I$ where I is given the discrete topology.

Preservation of Topological Properties

- Every disjoint union of discrete spaces is discrete
- *Separation*
- Every disjoint union of T_0 spaces is T_0
- Every disjoint union of T_1 spaces is T_1
- Every disjoint union of Hausdorff spaces is Hausdorff

- *Connectedness*
  - The disjoint union of two or more nonempty topological spaces is disconnected

Subspace Topology

In topology and related areas of mathematics, a subspace of a topological space X is a subset S of X which is equipped with a natural topology induced from that of X called the subspace topology (or the relative topology, or the induced topology, or the trace topology).

Definition

Given a topological space (X, τ) and a subset S of X, the subspace topology on is defined by

$$\tau_S = \{S \cap U \mid U \in \tau\}.$$

That is, a subset of S is open in the subspace topology if and only if it is the intersection of with an open set in (X, τ).

If S is equipped with the subspace topology then it is a topological space in its own right, and is called a subspace of (X, τ). Subsets of topological spaces are usually assumed to be equipped with the subspace topology unless otherwise stated.

Alternatively we can define the subspace topology for a subset S of X as the coarsest topology for which the inclusion map

$$\iota : S \to X$$

is continuous.

More generally, suppose i is an injection from a set S to a topological space X. Then the subspace topology on is defined as the coarsest topology for which is continuous.

The open sets in this topology are precisely the ones of the form for open in . is then homeomorphic to its image in (also with the subspace topology) and is called a topological embedding.

Examples

In the following, R represents the real numbers with their usual topology.

- The subspace topology of the natural numbers, as a subspace of R, is the discrete topology.
- The rational numbers Q considered as a subspace of R do not have the discrete topology (the point 0 for example is not an open set in Q). If a and b are rational, then the intervals (a, b) and [a, b] are respectively open and closed, but if a and b are irrational, then the set of all x with $a < x < b$ is both open and closed.
- The set [0,1] as a subspace of R is both open and closed, whereas as a subset of R it is only closed.
- As a subspace of R, $[0,1] \cup [2,3]$ is composed of two disjoint *open* subsets (which happen also to be closed), and is therefore a disconnected space.
- Let $S = [0,1)$ be a subspace of the real line R. Then [0,½) is open in S but not in R. Likewise [½, 1) is closed in S but not in R. S is both open and closed as a subset of itself but not as a subset of R.

Preservation of Topological Properties

If whenever a topological space has a certain topological property we have that all of its subspaces share the same property, then we say the topological property is hereditary. If only closed subspaces must share the property we call it weakly hereditary.

- Every open and every closed subspace of a topologically complete space is topologically complete.
- Every open subspace of a Baire space is a Baire space.
- Every closed subspace of a compact space is compact.
- Being a Hausdorff space is hereditary.
- Being a normal space is weakly hereditary.
- Total boundedness is hereditary.
- Being totally disconnected is hereditary.
- First countability and second countability are hereditary.

Quotient Space

In topology and related areas of mathematics, a quotient space (also called an identification space) is, intuitively speaking, the result of identifying or "gluing together" certain points of a given space. The points to be identified are specified by an equivalence relation. This is commonly done in order to construct new spaces from given ones.

Definition

Let (X,τ_X) be a topological space, and let ~ be an equivalence relation on X. The quotient space, $Y = X/\sim$ is defined to be the set of equivalence classes of elements of X:

$$Y = \{[x] : x \in X\} = \{\{v \in X : v \sim x\} : x \in X\},$$

equipped with the topology where the open sets are defined to be those sets of equivalence classes whose unions are open sets in X:

$$\tau_Y = \{U \subseteq Y : \bigcup U \in \tau_X\}.$$

Equivalently, we can define them to be those sets with an open preimage under the quotient map $q : X \to X/\sim$ which sends a point in X to the equivalence class containing it.

$$\tau_Y = \{U \subseteq Y : q^{-1}(U) \in \tau_X\}.$$

The quotient topology is the final topology on the quotient space with respect to the quotient map.

Examples

- Gluing. Often, topologists talk of gluing points together. If X is a topological space and points $x, y \in X$ are to be "glued", then what is meant is that we are to consider the quotient space obtained from the equivalence relation $a \sim b$ if and only if $a = b$ or $a = x$, $b = y$ (or $a = y$, $b = x$). The two points are henceforth interpreted as one point. Such gluing is commonly referred to as the wedge sum.
- Consider the unit square $I^2 = [0,1]\times[0,1]$ and the equivalence relation ~ generated by the requirement that all boundary points be equivalent, thus identifying all boundary points to a single equivalence class. Then $I^2/\sim$ is homeomorphic to the unit sphere S^2.
- Adjunction space. More generally, suppose X is a space and A is a subspace of X. One can identify all points in A to a single equivalence class and leave points

outside of A equivalent only to themselves. The resulting quotient space is denoted X/A. The 2-sphere is then homeomorphic to the unit disc with its boundary identified to a single point: $D^2/\mathrm{d}D^2$.

- Consider the set X = R of all real numbers with the ordinary topology, and write $x \sim y$ if and only if $x\mathrm{d}y$ is an integer. Then the quotient space $X/\sim$ is homeomorphic to the unit circle S^1 via the homeomorphism which sends the equivalence class of x to $\exp(2\pi ix)$.
- A vast generalization of the previous example is the following: Suppose a topological group G acts continuously on a space X. One can form an equivalence relation on X by saying points are equivalent if and only if they lie in the same orbit. The quotient space under this relation is called the orbit space, denoted X/G. In the previous example G = Z acts on R by translation. The orbit space R/Z is homeomorphic to S^1.

Warning: The notation R/Z is somewhat ambiguous. If Z is understood to be a group acting on R then the quotient is the circle. However, if Z is thought of as a subspace of R, then the quotient is an infinite bouquet of circles joined at a single point.

Properties

Quotient maps $q : X \rightarrow Y$ are characterised among surjective maps by the following property: if Z is any topological space and $f : Y \rightarrow Z$ is any function, then f is continuous if and only if $f \neq q$ is continuous.

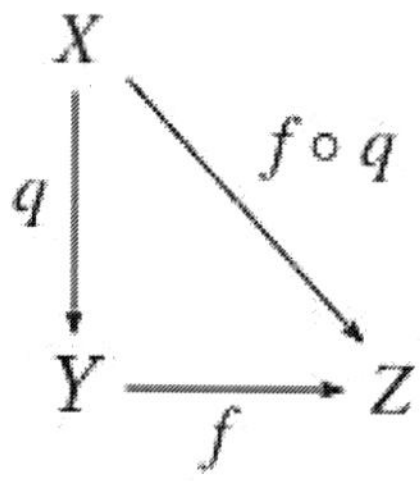

The quotient space $X/\sim$ together with the quotient map q : $X \rightarrow X/\sim$ is characterised by the following universal property: if $g : X \rightarrow Z$ is a continuous map such that $a \sim b$ implies $g(a) = g(b)$ for all a and b in X, then there exists a unique continuous map $f : X/\sim \to Z$ such that $g = f \neq q$. We say that g *descends to the quotient.*

The continuous maps defined on $X/\sim$ are therefore precisely those maps which arise from continuous maps defined on X that respect the equivalence relation (in the sense that they send equivalent elements to the same image).

This criterion is constantly used when studying quotient spaces.

Given a continuous surjection $f : X \to Y$ it is useful to have criteria by which one can determine if f is a quotient map.

Two sufficient criteria are that f be open or closed. Note that these conditions are only sufficient, not necessary. It is easy to construct examples of quotient maps that are neither open nor closed.

Compatibility with Other Topological Notions

- Separation
  - In general, quotient spaces are ill-behaved with respect to separation axioms. The separation properties of X need not be inherited by $X/\sim$, and $X/\sim$ may have separation properties not shared by X.
  - $X/\sim$ is a T1 space if and only if every equivalence class of $\sim$ is closed in X.
  - If the quotient map is open, then $X/\sim$ is a Hausdorff space if and only if $\sim$ is a closed subset of the product space $X \times X$.
- Connectedness
  - If a space is connected or path connected, then so are all its quotient spaces.
  - A quotient space of a simply connected or contractible space need not share those properties.
- Compactness
  - If a space is compact, then so are all its quotient spaces.
  - A quotient space of a locally compact space need not be locally compact.
- Dimension

— The topological dimension of a quotient space can be more (as well as less) than the dimension of the original space; space-filling curves provide such examples.

Relatively Compact Subspace

In mathematics, a relatively compact subspace (or relatively compact subset) Y of a topological space X is a subset whose closure is compact. Since closed subsets of a compact space are compact, every subset of a compact space is relatively compact.

In the case of a metric topology, or more generally when sequences may be used to test for compactness, the criterion for relative compactness becomes that any sequence in Y has a subsequence convergent in X.

Such a subset may also be called relatively bounded, or pre-compact, although the latter term is also used for a totally bounded subset. (These are equivalent in a complete space.)

Some major theorems characterise relatively compact subsets, in particular in function spaces. An example is the Arzelà–Ascoli theorem. Other cases of interest relate to uniform integrability, and the concept of normal family in complex analysis

.Mahler's compactness theorem in the geometry of numbers characterises relatively compact subsets in certain non-compact homogeneous spaces (specifically spaces of lattices).

The definition of an almost periodic function F at a conceptual level has to do with the translates of F being a relatively compact set. This needs to be made precise in terms of the topology used, in a particular theory.

Cohomotopy Group

In mathematics, particularly algebraic topology, cohomotopy sets are particular contravariant functors from the category of pointed topological spaces and point-preserving continuous maps to the category of sets and functions. They are dual to the homotopy groups, but less studied. The p-th cohomotopy set of a pointed topological space X is defined by

$$\pi^p(X) = [X, S^p]$$

the set of pointed homotopy classes of continuous mappings from X to the p-sphere S^p. For $p=1$ this set has an abelian group structure, and is isomorphic to the first cohomology group $H^1(X)$, since S^1 is a $K(\mathbb{Z},1)$. The set also has a group structure if X is a suspension ΣY, such as a sphere S^q for $q \geq 1$.

Brown's Representability Theorem

In mathematics, Brown's representability theorem in homotopy theory gives necessary and sufficient conditions on a contravariant functor F on the homotopy category *Hot* of pointed CW complexes, to the category of sets Set, to be a representable functor. That is, we are given

$$F\colon Hot^{op} \to \text{Set},$$

and certain necessary conditions for F to be of type $Hom(—, C)$ with C a CW-complex can be deduced from category theory alone. The statement of the substantive part of the theorem is that these necessary conditions are then sufficient. For technical reasons, the theorem is often stated for functors to the category of pointed sets; in other words the sets are also given a base point.

Brown Representability Theorem for CW Complexes

The representability theorem for CW complexes, due to E. H. Brown, is the following: suppose the functor F maps any colimit in *Hot* to a limit in Set. Then F is representable by some CW complex C, that is to say there is an isomorphism

$$F : Hot^{op} \to \mathbf{Set},$$

for any CW complex Z. This isomorphism is natural in Z in that for any morphism from Z to another CW complex Y the induced maps $F(Y) \to F(Z)$ and $Hom_{Hot}(Y, C) \to Hom_{Hot}(Z, C)$ are compatible with these isomorphisms. According to a combinatorial result in category theory, all small (that is to say, indexed over a set, as opposed to a proper class) colimits are built up from coproducts and pushouts (or just coequalisers). The traditional, equivalent statement of the theorem in algebraic topology is thus the following: suppose F satisfies the wedge axiom:

$$F(\vee_\alpha X_\alpha) \cong \prod_\alpha F(X_\alpha),$$

i.e., F converts any wedge sum (coproduct of pointed spaces) into a product of sets, and the Mayer-Vietoris axiom, requiring that for any CW complex W covered by two subcomplexes U and V, and any elements $u \in F(U), v \in F(V)$ such that u and v restrict to the same element of $F(U \cap V)$, there is an element $w \in F(W)$ restricting to u and v, respectively. Under these two conditions, F is representable.

The converse statement also holds: any functor represented by a CW complex satisfies these properties. This direction is an immediate consequence of basic homotopy theory. The deeper and more interesting part of the equivalence is the other implication.

Taking $F(X)$ to be the singular cohomology group $H^i(X,A)$ with coefficients in a given abelian group A, for fixed $i > 0$; then the representing space for F is the Eilenberg-MacLane space $K(A, i)$. This gives a means of showing the existence of Eilenberg-MacLane spaces.

Variants

A similar statement holds for spectra instead of CW complexes. The representing object C above can be shown to depend functorially on F: any natural transformation from F to another functor satisfying the conditions of the theorem necessarily induces a map of the representing objects. This is a consequence of Yoneda's lemma.

A version of the representability theorem in the case of triangulated categories is due to Amnon Neeman. Together with the preceding remark, it gives a criterion for a (covariant) functor F: $C \to D$ between triangulated categories satisfying certain technical conditions to have a right adjoint functor. Namely, if C and D are triangulated categories with C compactly generated and F a triangulated functor commuting with arbitrary direct sums, then F is a left adjoint. Neeman has applied this to proving the Grothendieck duality theorem in algebraic geometry. Jacob Lurie has proved a version of the Brown representability theorem for the homotopy category of a pointed quasicategory with a compact set of generators which are cogroup objects in the homotopy category. For instance, this applies to the homotopy category of (pointed) connected CW complexes, as well as to the unbounded derived category of a Grothendieck abelian category (in view of Lurie's higher-categorical refinement of the derived category).

Eilenberg–MacLane Space

In mathematics, an Eilenberg–MacLane space is a special kind of topological space that can be regarded as a building block for homotopy theory.

These spaces are important in many contexts in algebraic topology, including constructions of spaces, computations of homotopy groups of spheres, and definition of cohomology operations. The name is for Samuel Eilenberg and Saunders Mac Lane, who introduced such spaces in the late 1940s.

Let G be a group and n a positive integer. A connected topological space X is called an Eilenberg–MacLane space of type $K(G,n)$, if it has n-th homotopy group $\pi_n(X)$ isomorphic to G and all other homotopy groups trivial. If $n > 1$ then G must be abelian.

Then an Eilenberg–MacLane space exists, as a CW-complex, and is unique up to a weak homotopy equivalence. By abuse of language, any such space is often called just $K(G,n)$.

Examples

- The unit circle S^1 is a $K(Z,1)$.
- The infinite-dimensional complex projective space $P^{\infty}(C)$ is a model of $K(Z,2)$. This is one of the rare examples of classifying spaces admitting a manifold model, and is also the topological space the homotopy groups of which satisfy $\pi_i = 0$ for $i = 1$ and $i > 2$, while $\pi_2 = Z$. Its cohomology ring is $Z[x]$, namely the free polynomial ring on a single 2-dimensional generator $x \in H^2$. The generator can be represented in de Rham cohomology by the Fubini–Study 2-form. An application of $K(Z,2)$ is described at Abstract nonsense.
- The infinite-dimensional real projective space $P^{\infty}(R)$ is a $K(Z_2,1)$.
- The wedge sum of k unit circles $\vee_{i=1}^{k} S^1$ is a $K(G,1)$ for G the free group on k generators.

- The complement to any knot in a 3-dimensional sphere S^3 is of type $K(G,1)$; this is called the "asphericity of knots", and is a 1957 theorem of Christos Papakyriakopoulos.

Some further elementary examples can be constructed from these by using the obvious fact that the product $K(G,n) \times K(H,n)$ is $K(G \times H,n)$.

A $K(G,n)$ can be constructed stage-by-stage, as a CW complex, starting with a wedge of n-spheres, one for each generator of the group G, and adding cells in (possibly infinite number of) higher dimensions so as to kill all extra homotopy.

Properties of Eilenberg–MacLane Spaces

An important property of $K(G,n)$ is that, for any abelian group G, and any CW-complex X, the set

$$[X, K(G,n)]$$

of homotopy classes of maps from X to $K(G,n)$ is in natural bijection with the n-th singular cohomology group

$$H^n(X; G)$$

of the space X. Thus one says that the $K(G,n)$ are representing spaces for cohomology with coefficients in G. Since

$$H^n(K(G,n);G) = \mathrm{Hom}(H_n(K(G,n);\mathbb{Z}),G)$$
$$= \mathrm{Hom}(\pi_n(K(G,n)),G) = \mathrm{Hom}(G,G)$$

there is a distinguished element $u \in H^n(K(G,n);G)$ corresponding to the identity. The above bijection is given by pullback of that element — $f \mapsto f^*u$.

Another version of this result, due to Peter J. Huber, establishes a bijection with the n-th Èech cohomology group when X is Hausdorff and paracompact and G is countable, or when X is Hausdorff, paracompact and compactly generated and G is arbitrary.

A further result of Morita establishes a bijection with the n-th numerable Èech cohomology group for an arbitrary topological space X and G an arbitrary abelian group.

Every CW-complex possesses a Postnikov tower, that is, it is homotopy equivalent to an iterated fibration with fibres the Eilenberg–MacLane spaces.

There is a method due to Jean-Pierre Serre which allows one, at least theoretically, to compute homotopy groups of spaces using a spectral sequence for special fibrations with Eilenberg–MacLane spaces for fibres. The cohomology groups of Eilenberg–MacLane spaces can be used to classify all cohomology operations.

Fibre Bundle

In mathematics, and particularly topology, a fibre bundle (or, in British English, fibre bundle) is intuitively a space which *locally* "looks" like a certain product space, but *globally* may have a different topological structure. Specifically, the similarity between the fibre bundle E and a product space $B \times F$ is defined using a continuous surjective map

$$\pi : E \to B$$

that in small regions of E behaves just like a projection from corresponding regions of $B \times F$ to B. The map π, called the projection or submersion of the bundle, is regarded as part of the structure of the bundle. The space E is known as the total space of the fibre bundle, B as the base space, and F the fibre.

In the *trivial* case, E is just $B \times F$, and the map π is just the projection from the product space to the first factor. This is called a trivial bundle.

Examples of non-trivial fibre bundles, that is, bundles twisted in the large, include the Möbius strip and Klein bottle, as well as nontrivial covering spaces. Fibre bundles such as the tangent bundle of a manifold and more general vector bundles play an important role in differential geometry and differential topology, as do principal bundles.

Mappings which factor over the projection map are known as bundle maps, and the set of fibre bundles forms a category with respect to such mappings. A bundle map from the base space itself (with the identity mapping as projection) to E is called a section of E. Fibre bundles can be generalized in a number of ways, the most common of which is requiring that the transition between the local trivial patches should lie in a certain topological group, known as the structure group, acting on the fibre F.

Formal Definition

A fibre bundle consists of the data (E, B, π, F), where E, B, and F are topological spaces and $\pi : E \rightarrow B$ is a continuous surjection satisfying a *local triviality* condition outlined below.

The space B is called the base space of the bundle, E the total space, and F the fibre. The map π is called the projection map (or bundle projection). We shall assume in what follows that the base space B is connected. We require that for every x in E, there is an open neighbourhood $U \subset B$ of $\pi(x)$ (which will be called a trivializing neighbourhood) such that $\pi^{-1}(U)$ is homeomorphic to the product space $U \times F$, in such a way that π carries over to the projection onto the first factor. That is, the following diagram should commute:

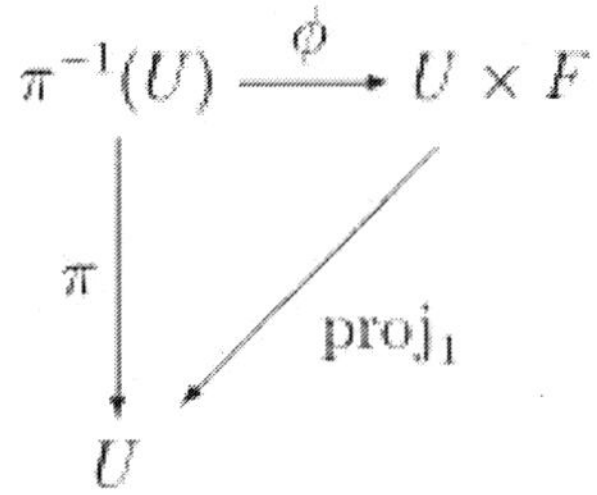

where $\mathrm{proj}_1 : U \times F \rightarrow U$ is the natural projection and $\varphi : \pi^{-1}(U) \rightarrow U \times F$ is a homeomorphism. The set of all $\{(U_i, \varphi_i)\}$ is called a local trivialization of the bundle.

Thus for any p in B, the preimage $\pi^{-1}(\{p\})$ is homeomorphic to F (since $\mathrm{proj}_1^{-1}(\{p\})$ clearly is) and is called the fibre over p. Every fibre bundle $\pi : E \rightarrow B$ is an open map, since projections of products are open maps. Therefore B carries the quotient topology determined by the map π.

A fibre bundle (E, B, π, F) is often denoted

$$F \longrightarrow E \xrightarrow{\ \pi\ } B$$

that, in analogy with a short exact sequence, indicates which space is the fibre, total space and base space, as well as the map from total to base space.

A smooth fibre bundle is a fibre bundle in the category of smooth manifolds. That is, E, B, and F are required to be smooth manifolds and all the functions above are required to be smooth maps.

Examples

Trivial Bundle

Let $E = B \times F$ and let $\pi : E \to B$ be the projection onto the first factor. Then E is a fibre bundle (of F) over B. Here E is not just locally a product but *globally* one. Any such fibre bundle is called a trivial bundle. Any fibre bundle over a contractible CW-complex is trivial.

Möbius Strip

Perhaps the simplest example of a nontrivial bundle E is the Möbius strip. It has the circle that runs lengthwise along the centre of the strip as a base B and a line segment for the fibre F, so the Möbius strip is a bundle of the line segment over the circle. A neighbourhood U of a point $x \in B$ is an arc; in the picture, this is the length of one of the squares. The preimage $\pi^{-1}(U)$ in the picture is a (somewhat twisted) slice of the strip four squares wide and one long. The homeomorphism φ maps the preimage of U to a slice of a cylinder: curved, but not twisted.

The corresponding trivial bundle $B \times F$ would be a cylinder, but the Möbius strip has an overall "twist". Note that this twist is visible only globally; locally the Möbius strip and the cylinder are identical (making a single vertical cut in either gives the same space).

Klein Bottle

A similar nontrivial bundle is the Klein bottle which can be viewed as a "twisted" circle bundle over another circle. The corresponding non-twisted (trivial) bundle is the 2-torus, $S^1 \times S^1$.

Covering Map

A covering space is a fibre bundle such that the bundle projection is a local homeomorphism. It follows in particular, that the fibre is a discrete space.

Vector and Principal Bundles

A special class of fibre bundles, called vector bundles, are those whose fibres are vector spaces (to qualify as a vector bundle the structure group of the bundle — must be a linear group). Important examples of vector bundles include the tangent bundle and cotangent bundle of a smooth manifold. From any vector bundle, one can construct the frame bundle of bases which is a principal bundle.

Another special class of fibre bundles, called principal bundles, are bundles on whose fibres a free and transitive action by a group G is given, so that each fibre is a principal homogeneous space. The bundle is often specified along with the group by referring to it as a principal G-bundle. The group G is also the structure group of the bundle. Given a representation ρ of G on a vector space V, a vector bundle with $\rho(G) \subseteq \mathrm{Aut}(V)$ as a structure group may be constructed, known as the associated bundle.

Sphere Bundles

A sphere bundle is a fibre bundle whose fibre is an n-sphere. Given a vector bundle E with a metric (such as the tangent bundle to a Riemannian manifold) one can construct the associated unit sphere bundle, for which the fibre over a point x is the set of all unit vectors in E_x. When the vector bundle in question is the tangent bundle T(M), the unit sphere bundle is known as the unit tangent bundle, and is denoted UT(M).

A sphere bundle is partially characterised by its Euler class, which is a degree n+1 cohomology class in the total space of the bundle. In the case n=1 the sphere bundle is called a circle bundle and the Euler class is equal to the first Chern class, which characterises the topology of the bundle completely. For any n, given the Euler class of a bundle, one can calculate its cohomology using a long exact sequence called the Gysin sequence.

Mapping Tori

If X is a topological space and $f:X \to X$ is a homeomorphism then the mapping torus M_f has a natural structure of a fibre bundle over the circle with fibre X. Mapping tori of homeomorphisms of surfaces are of particular importance in 3-manifold topology.

Quotient Spaces

If G is a topological group and H is a closed subgroup, then under some circumstances, the quotient space G/H together with the quotient map $\pi : G \to G/H$ is a fibre bundle, whose fibre is the topological space H. A necessary and sufficient condition for $(G,G/H,\pi,H)$ to form a fibre bundle is that the mapping π admit local cross-sections (Steenrod & 1951 §7).

The most general conditions under which the quotient map will admit local cross-sections are not known, although if G is a Lie group and H a closed subgroup (and thus a Lie subgroup by Cartan's theorem), then the quotient map is a fibre bundle. One example of this is the Hopf fibration, $S^3 \to S^2$ which is a fibre bundle over the sphere S^2 whose total space is S^3. From the perspective of Lie groups, S^3 can be identified with the special unitary group SU(2). The abelian subgroup of diagonal matrices is isomorphic to the circle group U(1), and the quotient SU(2)/U(1) is diffeomorphic to the sphere.

More generally, if G is any topological group and H a closed subgroup which also happens to be a Lie group, then $G \to G/H$ is a fibre bundle.

Sections

A section (or cross section) of a fibre bundle is a continuous map $f : B \to E$ such that $\pi(f(x))=x$ for all x in B. Since bundles do not in general have globally defined sections, one of the purposes of the theory is to account for their existence. The obstruction to the existence of a section can often be measured by a cohomology class, which leads to the theory of characteristic classes in algebraic topology.

The most well-known example is the hairy ball theorem, where the Euler class is the obstruction to the tangent bundle of the 2-sphere having a nowhere vanishing section.

Often one would like to define sections only locally (especially when global sections do not exist). A local section of a fibre bundle is a continuous map $f : U \to E$ where U is an open set in B and $\pi(f(x))=x$ for all x in U. If (U, φ) is a local trivialization chart then local sections always exist over U. Such sections are in 1-1 correspondence with continuous maps $U \to F$. Sections form a sheaf.

Differentiable Fibre Bundles

In the category of differentiable manifolds, fibre bundles arise naturally as submersions of one manifold to another. Not every (differentiable) submersion $f : M \to N$ from a differentiable manifold M to another differentiable manifold N gives rise to a differentiable fibre bundle. For one thing, the map must be surjective. However, this necessary condition is not quite sufficient, and there are a variety of sufficient conditions in common use.

If M and N are compact and connected, then any submersion $f : M \to N$ gives rise to a fibre bundle in the sense that there is a fibre space F diffeomorphic to each of the fibres such that $(E,B,\pi,F) = (M,N,f,F)$ is a fibre bundle. (Surjectivity of f follows by the assumptions already given in this case.) More generally, the assumption of compactness can be relaxed if the submersion $f : M \to N$ is assumed to be a surjective proper map, meaning that $f^{-1}(K)$ is compact for every compact subset K of N. Another sufficient condition, due to Ehresmann (1951), is that if $f : M \to N$ is a surjective submersion with M and N differentiable manifolds such that the preimage $f^{-1}\{x\}$ is compact and connected for all $x - N$, then f admits a compatible fibre bundle structure (Michor 2008, §17).

Generalizations

- The notion of a bundle applies to many more categories in mathematics, at the expense of appropriately modifying the local triviality condition.
- In topology, a fibration is a mapping $\pi : E \to B$ which has certain homotopy-theoretic properties in common with fibre bundles. Specifically, under mild technical assumptions a fibre bundle always has the homotopy lifting property or homotopy covering property. This is the defining property of a fibration.

Möbius Strip

The Möbius strip or Möbius band, also Mobius or Moebius, is a surface with only one side and only one boundary component. The Möbius strip has the mathematical property of being non-orientable. It can be realised as a ruled surface. It was discovered independently by the German mathematicians August Ferdinand Möbius and Johann Benedict Listing in 1858.

A model can easily be created by taking a paper strip and giving it a half-twist, and then joining the ends of the strip together to form a loop. In Euclidean space there are two types of Möbius strips depending on the direction of the half-twist: clockwise and counterclockwise. That is to say, it is a chiral object with "handedness" (right-handed or left-handed).

It is straightforward to find algebraic equations the solutions of which have the topology of a Möbius strip, but in general these equations do not describe the same geometric shape that one gets from the twisted paper model described above. In particular, the twisted paper model is a developable surface (it has zero Gaussian curvature). A system of differential-algebraic equations that describes models of this type was published in 2007 together with its numerical solution.

The Euler characteristic of the Möbius strip is zero.

Properties

The Möbius strip has several curious properties. A line drawn starting from the seam down the middle will meet back at the seam but at the "other side". If continued the line will meet the starting point and will be double the length of the original strip. This single continuous curve demonstrates that the Möbius strip has only one boundary.

Cutting a Möbius strip along the centre line with a pair of scissors yields one long strip with two full twists in it, rather than two separate strips; the result is not a Möbius strip. This happens because the original strip only has one edge that is twice as long as the original strip. Cutting creates a second independent edge, half of which was on each side of the scissors. Cutting this new, longer, strip down the middle creates two strips wound around each other, each with two full twists. If the strip is cut along about a third of the way in from the edge, it creates two strips: One is a thinner Möbius strip — it is the centre third of the original strip, comprising 1/3 of the width and the same length as the original strip. The other is a longer but thin strip with two full twists in it — this is a neighbourhood of the edge of the original strip, and it comprises 1/3 of the width and twice the length of the original strip.

Other analogous strips can be obtained by similarly joining strips with two or more half-twists in them instead of one. For example, a strip with three half-twists, when divided lengthwise, becomes a strip tied in a trefoil knot. (If this knot is unravelled, the strip is made with eight half-twists in addition to an overhand knot.) A strip with N half-twists, when bisected, becomes a strip with $N + 1$ full twists. Giving it extra twists and reconnecting the ends produces figures called paradromic rings. A strip with an odd-number of half-twists, such as the Möbius strip, will have only one surface and one boundary. A strip twisted an even number of times will have two surfaces and two boundaries.

If a strip with an odd number of half-twists is cut in half along its length, it will result in a longer strip, with the same number of loops as there are half-twists in the original. Alternatively, if a strip with an even number of half-twists is cut in half along its length, it will result in two linked strips, each with the same number of twists as the original.

Geometry and Topology

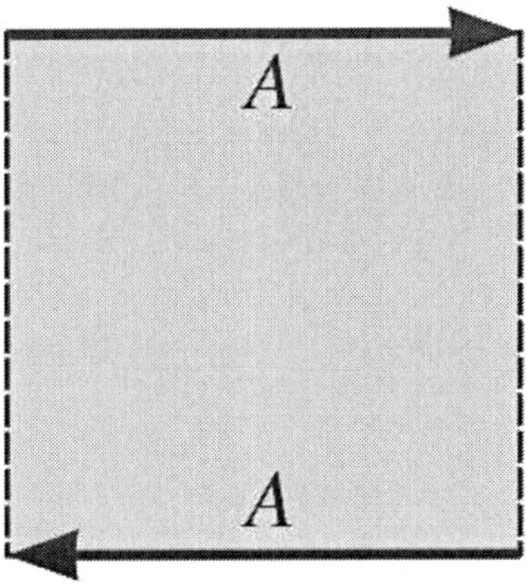

Figure: *To turn a rectangle into a Möbius strip, join the edges labelled A so that the directions of the arrows match.*

One way to represent the Möbius strip as a subset of R^3 is using the parametrization:

$$x(u,v) = \left(1 + \frac{1}{2} v \cos\frac{u}{2}\right) \cos u$$

$$y(u,v) = \left(1 + \frac{1}{2} v \cos\frac{u}{2}\right) \sin u$$

$$z(u,v) = \frac{1}{2} v \sin\frac{u}{2}$$

where $0 \leq u < 2\pi$ and $^-1 \leq v \leq 1$. This creates a Möbius strip of width 1 whose centre circle has radius 1, lies in the *xy* plane and is centred at (0, 0, 0). The parameter *u* runs around the strip while *v* moves from one edge to the other.

In cylindrical polar coordinates (r, θ, z), an unbounded version of the Möbius strip can be represented by the equation:

$$\log(r)\sin\left(\frac{1}{2}\theta\right) = z\cos\left(\frac{1}{2}\theta\right).$$

Topologically, the Möbius strip can be defined as the square $[0,1] \times [0,1]$ with its top and bottom sides identified by the relation $(x, 0) \sim (1 - x, 1)$ for $0 \leq x \leq 1$, as in the diagram on the right.

A less used presentation of the Möbius strip is as the orbifold quotient of a torus. A torus can be constructed as the square $[0,1] \times [0,1]$ with the edges identified as $(0,y) \sim (1,y)$ (glue left to right) and $(x,0) \sim (x,1)$ (glue bottom to top). If one then also identified $(x,y) \sim (y,x)$, then one obtains the Möbius strip. The diagonal of the square (the points (x,x) where both coordinates agree) becomes the boundary of the Möbius strip, and carries an orbifold structure, which geometrically corresponds to "reflection" – geodesics (straight lines) in the Möbius strip reflect off the edge back into the strip. Notationally, this is written as T^2/S_2 – the 2-torus quotiented by the group action of the symmetric group on two letters (switching coordinates), and it can be thought of as the configuration space of two unordered points on the circle, possibly the same (the edge corresponds to the points being the same), with the torus corresponding to two ordered points on the circle.

The Möbius strip is a two-dimensional compact manifold (i.e. a surface) with boundary. It is a standard example of a surface which is not orientable. The Möbius strip is also a standard example used to illustrate the mathematical concept of a fibre bundle. Specifically, it is a nontrivial bundle over the circle S^1 with a fibre the unit interval, $I = [0,1]$. Looking only at the edge of the Möbius strip gives a nontrivial two point (or Z_2) bundle over S^1. A simple construction of the Möbius strip which can be used to portray it in computer graphics or modelling packages is as follows :

- Take a rectangular strip. Rotate it around a fixed point not in its plane. At every step also rotate the strip along a line in its plane (the line which divides the strip in two) and perpendicular to the main orbital radius. The surface generated on one complete revolution is the Möbius strip.
- Take a Möbius strip and cut it along the middle of the strip. This will form a new strip, which is a rectangle joined by rotating one end a whole turn. By cutting it down the middle again, this forms two interlocking whole-turn strips.

Related Objects

A closely related 'strange' geometrical object is the Klein bottle. A Klein bottle can be produced by gluing two Möbius strips together along their edges; this cannot be done in ordinary three-dimensional Euclidean space without creating self-intersections.

Another closely related manifold is the real projective plane. If a circular disk is cut out of the real projective plane, what is left is a Möbius strip. Going in the other direction, if one glues a disk to a Möbius strip by identifying their boundaries, the result is the projective plane. In order to visualize this, it is helpful to deform the Möbius strip so that its boundary is an ordinary circle. The real projective plane, like the Klein bottle, cannot be embedded in three-dimensions without self-intersections.

In graph theory, the Möbius ladder is a cubic graph closely related to the Möbius strip.

In 1968, Gonzalo Vélez Jahn (UCV, Caracas, Venezuela) discovered three dimensional bodies with Möbian characteristics, later described by Martin Gardner as prismatic rings that became 'toroidal polyhedrons'.

Applications

There have been several technical applications for the Möbius strip. Giant Möbius strips have been used as conveyor belts that last longer because the entire surface area of the belt gets the same amount of wear, and as continuous-loop recording tapes (to double the playing time). Möbius strips are common in the manufacture of fabric computer printer and typewriter ribbons, as they allow the ribbon to be twice as wide as the print head while using both halves evenly. A Möbius resistor is an electronic circuit element that cancels its own inductive reactance. Nikola Tesla patented similar technology in 1894: "Coil for Electro Magnets" was intended for use with his system of global transmission of electricity without wires.

The Möbius strip is the configuration space of two unordered points on a circle. Consequently, in music theory, the space of all two note chords, known as dyads, takes the shape of a Möbius strip; this and generalizations to more points is a significant application of orbifolds to music theory.

In physics/electro-technology:

- As a compact resonator with the resonance frequency which is half that of identically constructed linear coils
- As an inductionless resistor
- As superconductors with high transition temperature

In chemistry/nano-technology:

- As molecular knots with special characteristics (Knotane, Chirality)
- As molecular engines
- As graphene volume (nano-graphite) with new electronic characteristics, like helical magnetism
- In a special type of aromaticity: Möbius aromaticity
- Charged particles that have been caught in the magnetic field of the earth can move on a Möbius band
- The cyclotide (cyclic protein) Kalata B1, active substance of the plant Oldenlandia affinis, contains Möbius topology for the peptide backbone.

Line Bundle

In mathematics, a line bundle expresses the concept of a line that varies from point to point of a space. For example a curve in the plane having a tangent line at each point determines a varying line: the tangent bundle is a way of organising these. More formally, in algebraic topology and differential topology a line bundle is defined as a vector bundle of rank 1.

One-dimensional real line bundles (as just described) and one-dimensional complex line bundles differ. The topology of the 1×1 invertible real matrices is a space homotopy equivalent to a discrete two-point space (positive and negative reals contracted down), while 1×1 invertible complex matrices have the homotopy type of a circle. A real line bundle is therefore in the eyes of homotopy theory as good as a fibre bundle with a two-point fibre - a double covering.

This is like the orientable double cover on a differential manifold: indeed that's a special case in which the line bundle is the determinant bundle (top exterior power) of the tangent bundle. The Möbius strip corresponds to a double cover of the circle (the $\theta \to 2\theta$ mapping) and can be viewed if we wish as having fibre two points, the unit interval or the real line: the data are equivalent. In the case of the complex line bundle, we are looking in fact also for circle bundles. There are some celebrated ones, for example the Hopf fibrations of spheres to spheres.

Tautological Line Bundle

The tautological line bundle on a projective space appears frequently in mathematics, often in the study of characteristic classes. Note that some prominent topologists once called this the canonical line bundle. However, this terminology is now strongly deprecated, as it conflicts with universally accepted terminology the theory of complex manifolds, where the canonical line bundle of a complex m-manifold is defined to be the bundle of (m,0)-forms. For this reason, the term *tautological* is now widely considered to be preferrable, as it is universally acceptable, and avoids needless confusion.

Definition

Form the cartesian product $\mathbf{R}P^n \times \mathbf{R}^{n+1}$, with the first factor denoting real projective n-space. We consider the subset

$$E(\gamma_n) := \{(x,v) \in \mathbf{R}P^n \times \mathbf{R}^{n+1} : v \in x\}.$$

We have an obvious projection map $\pi : E(\gamma_n) \to \mathbf{R}P^n$, with $(x,v) \mapsto x$. Each fibre of π is then the line x inside Euclidean (n+1)-space. Giving each fibre the induced vector space structure we obtain the bundle

$$\gamma_n := (E(\gamma_n) \to \mathbf{R}P^n),$$

the tautological line bundle over $\mathbf{R}P^n$.

Complex and Quaternionic Cases

The above definition continues to makes sense if we replace the field R by either the complex numbers C or the quaternions Thus we obtain the complex line bundle

$$\gamma_{n,\mathbf{C}} := (E(\gamma_{n,\mathbf{C}}) \to \mathbf{C}P^n),$$

whose fibres are isomorphic to $\mathbf{C} \cong \mathbf{R}^2$, and the quaternionic line bundle

$$\gamma_{n,\mathbf{H}} := (E(\gamma_{n,\mathbf{H}}) \to \mathbf{H}P^n),$$

whose fibres are isomorphic to $\mathbf{H} \cong \mathbf{R}^4$.

Tautological Line Bundle in Algebraic Geometry

In algebraic geometry, this notion exists over any commutative unital ring.

Over a field, its dual line bundle is the line bundle associated to the hyperplane divisor *H*, whose global sections are the linear forms. Its Chern class is *-H*. This is an example of an anti-ample line bundle. Over $\mathbb{C}$, this is equivalent to saying that it is a negative line bundle, meaning that minus its Chern class is the de Rham class of a Kähler form.

Facts

- The tautological line bundle γ_n is locally trivial but not trivial, for $n \geq 1$. This remains true over other fields.

In fact, it is straightforward to show that, for $n = 1$, the real tautological line bundle is none other than the well-known bundle whose total space is the Möbius strip. For a full proof of the above fact, see.

Maps to Projective Space

Suppose that X is a space and that L is a line bundle on X. A global section of L is a function $s : X \to L$ such that if $p : L \to X$ is the natural projection, then $ps = \mathrm{id}_X$. In a small neighbourhood U in X in which L is trivial, the total space of the line bundle is the product of U and the underlying field k, and the section s restricts to a function $U \to k$. However, the values of s depend on the choice of trivialization, and so they are determined only up to multiplication by a nowhere-vanishing function.

Global sections determine maps to projective spaces in the following way: Choosing $r + 1$ not all zero points in a fibre of L chooses a fibre of the tautological line bundle on P^r, so choosing $r + 1$ non-simultaneously vanishing global sections of L determines a map from X into projective space P^r. This map sends the fibres of L to the fibres of the dual of the tautological bundle. More specifically, suppose that s_0, ..., s_r are global sections of L. In a small neighbourhood U in X, these sections determine k-valued functions on U whose values depend on the choice of trivialization. However, they are determined up to *simultaneous* multiplication by a non-zero function, so their ratios are well-defined.

That is, over a point x, the values $s_0(x)$, ..., $s_r(x)$ are not well-defined because a change in trivialization will multiply them each by a non-zero constant λ. But it will multiply them by the *same* constant λ, so the homogeneous coordinates $[s_0(x) : ... : s_r(x)]$ are well-defined as long as the sections s_0, ..., s_r do not simultaneously vanish at x.

Therefore, if the sections never simultaneously vanish, they determine a form $[s_0 : ... : s_r]$ which gives a map from X to P^r, and the pullback of the dual of the tautological bundle under this map is L. In this way, projective space acquires a universal property.

The universal way to determine a map to projective space is to map to the projectivization of the vector space of all sections of L. In the topological case, there is a non-vanishing section at every point which can be constructed using a bump function which vanishes outside a small neighbourhood of the point.

Because of this, the resulting map is defined everywhere. However, the codomain is usually far, far too big to be useful. The opposite is true in the algebraic and holomorphic settings. Here the space of global sections is often finite dimensional, but there may not be any non-vanishing global sections at a given point.

(As in the case when this procedure constructs a Lefschetz pencil.) In fact, it is possible for a bundle to have no non-zero global sections at all; this is the case for the tautological line bundle. When the line bundle is sufficiently ample this construction verifies the Kodaira embedding theorem.

Determinant Bundles

In general if V is a vector bundle on a space X, with constant fibre dimension n, the n-th exterior power of V taken fibre-by-fibre is a line bundle, called the determinant line bundle. This construction is in particular applied to the cotangent bundle of a smooth manifold.

The resulting determinant bundle is responsible for the phenomenon of tensor densities, in the sense that for an orientable manifold it has a global section, and its tensor powers with any real exponent may be defined and used to 'twist' any vector bundle by tensor product.

Characteristic Classes, Universal Bundles and Classifying Spaces

The first Stiefel–Whitney class classifies smooth real line bundles; in particular, the collection of (equivalence classes of) real line bundles are in correspondence with elements of the first cohomology with Z/2Z coefficients; this correspondence is in fact an isomorphism of abelian groups (the group operations being tensor product of line bundles and the usual addition on cohomology). Analogously, the first Chern class classifies smooth complex line bundles on a space, and the group of line bundles is isomorphic to the second cohomology class with integer coefficients. However, bundles can have equivalent smooth structures (and thus the same first Chern class) but different holomorphic structures. The Chern class statements are easily proven using the exponential sequence of sheaves on the manifold.

One can more generally view the classification problem from a homotopy-theoretic point of view. There is a universal bundle for real line bundles, and a universal bundle for complex line bundles. According to general theory about classifying spaces, the heuristic is to look for contractible spaces on which there are group actions of the respective groups C_2 and S^1, that are free actions. Those spaces can serve as the universal principal bundles, and the quotients for the actions as the classifying spaces *BG*. In these cases we can find those explicitly, in the infinite-dimensional analogues of real and complex projective space.

Therefore the classifying space BC_2 is of the homotopy type of RP", the real projective space given by an infinite sequence of homogeneous coordinates. It carries the universal real line bundle; in terms of homotopy theory that means that any real line bundle *L* on a CW complex *X* determines a *classifying map* from *X* to RP", making *L* a bundle isomorphic to the pullback of the universal bundle. This classifying map can be used to define the Stiefel-Whitney class of *L*, in the first cohomology of *X* with Z/2Z coefficients, from a standard class on RP".

In an analogous way, the complex projective space CP carries a universal complex line bundle. In this case classifying maps give rise to the first Chern class of *X*, in $H^2(X)$ (integral cohomology).

There is a further, analogous theory with quaternionic (real dimension four) line bundles. This gives rise to one of the Pontryagin classes, in real four-dimensional cohomology. In this way foundational cases for the theory of characteristic classes depend only on line bundles. According to a general splitting principle this can determine the rest of the theory (if not explicitly).

There are theories of holomorphic line bundles on complex manifolds, and invertible sheaves in algebraic geometry, that work out a line bundle theory in those areas.

Hopf Fibration

In the mathematical field of topology, the Hopf fibration (also known as the Hopf bundle or Hopf map) describes a 3-sphere (a hypersphere in four-dimensional space) in terms of circles and an ordinary sphere. Discovered by Heinz Hopf in 1931, it is an influential early example of a fibre bundle. Technically, Hopf found a many-to-one continuous function (or "map") from the 3-sphere onto the 2-sphere such that each distinct *point* of the 2-sphere comes from a distinct *circle* of the 3-sphere (Hopf 1931). Thus the 3-sphere is composed of fibres, where each fibre is a circle — one for each point of the 2-sphere.

This fibre bundle structure is denoted

$$S^1 \to S^3 \xrightarrow{p} S^2,$$

meaning that the fibre space S^1 (a circle) is embedded in the total space S^3 (the 3-sphere), and p: $S^3 \to S^2$ (Hopf's map) projects S^3 onto the base space S^2 (the ordinary 2-sphere). The Hopf fibration, like any fibre bundle, has the important property that it is locally a product space. However it is not a *trivial* fibre bundle, i.e., S^3 is not *globally* a product of S^2 and S^1 although locally it is indistinguishable from it.

This has many implications: for example the existence of this bundle shows that the higher homotopy groups of spheres are not trivial in general. It also provides a basic example of a principal bundle, by identifying the fibre with the circle group. Stereographic projection of the Hopf fibration induces a remarkable structure on R^3, in which space is filled with nested tori made of linking Villarceau circles. Here each fibre projects to a circle in space (one of which is a "circle through infinity" — a line). Each torus is the stereographic projection of the inverse image of a circle of latitude of the 2-sphere. (Topologically, a torus is the product of two circles.) One of these tori is illustrated by the image of linking keyrings on the right.

There are numerous generalizations of the Hopf fibration. The unit sphere in C^{n+1} fibres naturally over CP^n with circles as fibres, and there are also real, quaternionic, and octonionic versions of these fibrations. In particular, the Hopf fibration belongs to a family of four fibre bundles in which the total space, base space, and fibre space are all spheres:

$$S^0 \to S^1 \to S^1,$$
$$S^1 \to S^3 \to S^2,$$
$$S^3 \to S^7 \to S^4,$$
$$S^7 \to S^{15} \to S^8.$$

By Adams' theorem such fibrations can occur only in these dimensions. The Hopf fibration is important in twistor theory.

Definition and Construction

For any natural number n, an n-dimensional sphere, or n-sphere, can be defined as the set of points in an $(n+1)$-

dimensional space which are a fixed distance from a central point. For concreteness, the central point can be taken to be the origin, and the distance of the points on the sphere from this origin can be assumed to be a unit length. With this convention, the n-sphere, S^n, consists of the points $(x_1, x_2, \ldots, x_{n+1})$ in R^{n+1} with $x\, x_1^2 + x_2^2 + \cdots + x_{n+1}^2 = 1$. For example, the 3-sphere consists of the points (x_1, x_2, x_3, x_4) in R^4 with $x_1^2 + x_2^2 + x_3^2 + x_4^2 = 1$.

The Hopf fibration p: $S^3 \rightarrow S^2$ of the 3-sphere over the 2-sphere can be defined in several ways.

Direct Construction

Identify R^4 with C^2 and R^3 with C×R (where C denotes the complex numbers) by writing:

(x_1, x_2, x_3, x_4) as $(z_0 = x_1 + ix_2, z_1 = x_3 + ix_4)$; and

(x_1, x_2, x_3) as $(z = x_1 + ix_2, x = x_3)$.

Thus S^3 is identified with the subset of all (z_0, z_1) in C^2 such that $|z_0|^2 + |z_1|^2 = 1$, and S^2 is identified with the subset of all (z, x) in C×R such that $|z|^2 + x^2 = 1$. (Here, for a complex number $z = x + iy$, $|z|^2 = z z^- = x^2 + y^2$, where the star denotes the complex conjugate.) Then the Hopf fibration p is defined by

$$p(z_0, z_1) = (2z_0z_1^-, |z_0|^2 - |z_1|^2).$$

The first component is a complex number, whereas the second component is real. Any point on the 3-sphere must have the property that $|z_0|^2 + |z_1|^2 = 1$.

If that is so, then $p(z_0, z_1)$ lies on the unit 2-sphere in C×R, as may be shown by squaring the complex and real components of p

$$2z_0z_1^* \cdot 2z_0^*z_1 + \left(|z_0|^2 - |z_1|^2\right)^2 = 4|z_0|^2|z_1|^2 + |z_0|^4 - 2|z_0|^2|z_1|^2 + |z_1|^4$$

$$= \left(|z_0|^2 + |z_1|^2\right)^2 = 1$$

Furthermore, if two points on the 3-sphere map to the same point on the 2-sphere, i.e., if $p(z_0, z_1) = p(w_0, w_1)$, then (w_0, w_1) must equal $(\lambda z_0, \lambda z_1)$ for some complex number λ with $|\lambda|^2 = 1$.

The converse is also true; any two points on the 3-sphere that differ by a common complex factor λ map to the same point on the 2-sphere.

These conclusions follow, because the complex factor λ cancels with its complex conjugate λ in both parts of p: in the complex $2z_0z_1^-$ component and in the real component $|z_0|^2 - |z_1|^2$.

Since the set of complex numbers λ with $|\lambda|^2 = 1$ form the unit circle in the complex plane, it follows that for each point m in S^2, the inverse image $p^{-1}(m)$ is a circle, i.e., $p^{-1}m$ E– S^1. Thus the 3-sphere is realised as a disjoint union of these circular fibres.

Geometric Interpretation Using Rotations

Another geometric interpretation of the Hopf fibration can be obtained by considering rotations of the 2-sphere in ordinary 3-dimensional space. The rotation group SO(3) has a double cover, the spin group Spin(3), diffeomorphic to the 3-sphere. The spin group acts transitively on S^2 by rotations. The stabilizer of a point is isomorphic to the circle group. It follows easily that the 3-sphere is a principal circle bundle over the 2-sphere, and this is the Hopf fibration.

To make this more explicit, there are two approaches: the group Spin(3) can either be identified with the group Sp(1) of unit quaternions, or with the special unitary group SU(2).

In the first approach, a vector (x_1, x_2, x_3, x_4) in R^4 is interpreted as a quaternion $q \in$ H by writing

$$q = x_1 + \mathbf{i}x_2 + \mathbf{j}x_3 + \mathbf{k}x_4.$$

The 3-sphere is then identified with the quaternions of unit norm, i.e., those $q \in \mathbf{H}$ for which $|q|^2 = 1$, where $|q|^2 = q\, q^*$, which is equal to $x_1^2 + x_2^2 + x_3^2 + x_4^2$ for q as above.

On the other hand, a vector (y_1, y_2, y_3) in R^3 can be interpreted as an imaginary quaternion

$$p = \mathbf{i}y_1 + \mathbf{j}y_2 + \mathbf{k}y_3.$$

Then, as is well-known since Cayley (1845), the mapping

$$p \mapsto qpq^*$$

is a rotation in R³: indeed it is clearly an isometry, since $|q\,p\,q^*|^2 = q\,p\,q^*q\,p^*q^* = q\,p\,p^*q^* = |p|^2$, and it is not hard to check that it preserves orientation.

In fact, this identifies the group of unit quaternions with the group of rotations of R³, modulo the fact that the unit quaternions q and ^-q determine the same rotation. As noted above, the rotations act transitively on S^2, and the set of unit quaternions q which fix a given unit imaginary quaternion p have the form $q = u + v\,p$, where u and v are real numbers with $u^2 + v^2 = 1$. This is a circle subgroup. For concreteness, one can take p = k, and then the Hopf fibration can be defined as the map sending a unit quaternion q to q k q^-.

This approach is related to the direct construction by identifying a quaternion $q = x_1 + \mathrm{i}\,x_2 + \mathrm{j}\,x_3 + \mathrm{k}\,x_4$ with the 2×2 matrix:

$$\begin{bmatrix} x_1 + \mathbf{i}x_2 & x_3 + \mathbf{i}x_4 \\ -x_3 + \mathbf{i}x_4 & x_1 - \mathbf{i}x_2 \end{bmatrix}.$$

This identifies the group of unit quaternions with SU(2), and the imaginary quaternions with the skew-hermitian 2×2 matrices (isomorphic to C×R).

Explicit Formulae

The rotation induced by a unit quaternion q = w + *i* x + *j* y + *k* z *is given explicitly by the orthogonal matrix*

$$\begin{bmatrix} 1-2(y^2+z^2) & 2(xy-wz) & 2(xz+wy) \\ 2(xy+wz) & 1-2(x^2+z^2) & 2(yz-wx) \\ 2(xz-wy) & 2(yz+wx) & 1-2(x^2+y^2) \end{bmatrix}.$$

Here we find an explicit real formula for the bundle projection. For, the fixed unit vector along the z axis, (0,0,1), rotates to another unit vector,

$$\left(2(xz+wy), 2(yz-wx), 1-2(x^2+y^2)\right),$$

which is a continuous function of (w,x,y,z). That is, the image of q is where it aims the z axis. The fibre for a given point on S^2 consists of all those unit quaternions that aim there. To write an explicit formula for the fibre over a point (a,b,c) in S^2, we may proceed as follows. Multiplication of unit quaternions produces composition of rotations, and

$$q_\theta = \cos\theta + \mathbf{k}\sin\theta$$

is a rotation by 2θ around the z axis. As θ varies, this sweeps out a great circle of S^3, our prototypical fibre. So long as the base point, (a,b,c), is not the antipode, (0,0,–1), the quaternion

$$q_{(a,b,c)} = \frac{1}{\sqrt{2(1+c)}}(1+c-\mathbf{i}b+\mathbf{j}a)$$

will aim there. Thus the fibre of (a,b,c) is given by quaternions of the form $q_{(a,b,c)}q_\theta$, which are the S^3 points

$$\frac{1}{\sqrt{2(1+c)}}\big((1+c)\cos(\theta), a\sin(\theta)$$

$$-b\cos(\theta), a\cos(\theta)+b\sin(\theta), (1+c)\sin(\theta)\big).$$

Since multiplication by $q_{(a,b,c)}$ acts as a rotation of quaternion space, the fibre is not merely a topological circle, it is a geometric circle. The final fibre, for (0,0,–1), can be given by using $q_{(0,0,-1)}$ = i, producing

$$\big(0, \cos(\theta), -\sin(\theta), 0\big),c$$

which completes the bundle.

Thus, a simple way of visualizing the Hopf fibration is as follows. Any point on the 3-sphere is equivalent to a quaternion, which in turn is equivalent to a particular rotation of a Cartesian coordinate frame in three dimensions.

The set of all possible quaternions produces the set of all possible rotations, which moves the tip of one unit vector of such a coordinate frame (say, the z vector) to all possible points on a unit 2-sphere. However, fixing the tip of the z vector does not specify the rotation fully; a further rotation is possible about the z-axis.

Thus, the 3-sphere is mapped onto the 2-sphere, plus a single rotation.

Fluid Mechanics

If the Hopf fibration is treated as a vector field in 3 dimensional space then there is a solution to the (compressible, non-viscous) Navier-Stokes equations of fluid dynamics in which the fluid flows along the circles of the projection of the Hopf fibration in 3 dimensional space. The size of the velocities, the density and the pressure can be chosen at each point to satisfy the equations. All these quantities fall to zero going away from the centre. If a is the distance to the inner ring, the velocities, pressure and density fields are given by:

$$\mathbf{v}(x,y,z) = A\left(a^2+x^2+y^2+z^2\right)^{-2}$$

$$\left(2(-ay+xz), 2(ax+yz), a^2-x^2-y^2+z^2\right)$$

$$p(x,y,z) = -A^2B\left(a^2+x^2+y^2+z^2\right)^{-3}, \rho(x,y,z)$$

$$= 3B\left(a^2+x^2+y^2+z^2\right)^{-1}$$

for arbitrary constants A and B. Similar patterns of fields are found as soliton solutions of magnetohydrodynamics :

Generalizations

The Hopf construction, viewed as a fibre bundle p: $S^3 \to CP^1$, admits several generalizations, which are also often known as Hopf fibrations. First, one can replace the projective line by an n-dimensional projective space. Second, one can replace the complex numbers by any (real) division algebra, including (for n = 1) the octonions.

Real Hopf Fibrations

A real version of the Hopf fibration is obtained by regarding the circle S^1 as a subset of R^2 in the usual way and by identifying antipodal points. This gives a fibre bundle $S^1 \to RP^1$ over the real projective line with fibre $S^0 = \{1, -1\}$. Just as CP^1 is diffeomorphic to a sphere, RP^1 is diffeomorphic to a circle.

More generally, the n-sphere S^n fibres over real projective space RP^n with fibre S^0.

Complex Hopf Fibrations

The Hopf construction gives circle bundles $p : S^{2n+1} \to CP^n$ over complex projective space. This is actually the restriction of the tautological line bundle over CP^n to the unit sphere in C^{n+1}.

Quaternionic Hopf Fibrations

Similarly, one can regard S^{4n-1} as lying in H^n (quaternionic n-space) and factor out by unit quaternion (= S^3) multiplication to get HP^n. In particular, since $S^4 = HP^1$, there is a bundle $S^7 \to S^4$ with fibre S^3.

Octonionic Hopf Fibrations

A similar construction with the octonions yields a bundle $S^{15} \to S^8$ with fibre S^7. One can regard S^8 as the octonionic projective line OP^1. Although one can also define an octonionic projective plane, OP^2, the sphere S^{31} does not fibre over it.

Fibrations between Spheres

Sometimes the term "Hopf fibration" is restricted to the fibrations between spheres obtained above, which are

- $S^1 \to S^1$ with fibre S^0
- $S^3 \to S^2$ with fibre S^1
- $S^7 \to S^4$ with fibre S^3
- $S^{15} \to S^8$ with fibre S^7

As a consequence of Adams' theorem, fibre bundles with spheres as total space, base space, and fibre can occur only in these dimensions. Fibre bundles with similar properties, but different from the Hopf fibrations, were used by John Milnor to construct exotic spheres.

Geometry and Applications

The Hopf fibration has many implications, some purely attractive, others deeper. For example, stereographic projection of S^3 to R^3 induces a remarkable structure in R^3, which in turn illuminates the topology of the bundle (Lyons 2003). Stereographic projection preserves circles and maps the Hopf fibres to geometrically perfect circles in R^3 which fill space. Here there is one exception: the Hopf circle containing the projection point maps to a straight line in R^3 — a "circle through infinity".

The fibres over a circle of latitude on S^2 form a torus in S^3 (topologically, a torus is the product of two circles) and these project to nested toruses in R^3 which also fill space. The individual fibres map to linking Villarceau circles on these tori, with the exception of the circle through the projection point and the one through its opposite point: the former maps to a straight line, the latter to a unit circle perpendicular to, and centred on, this line, which may be viewed as a degenerate torus whose radius has shrunken to zero. Every other fibre image encircles the line as well, and so, by symmetry, each circle is linked through *every* circle, both in R^3 and in S^3. Two such linking circles form a Hopf link in R^3

Hopf proved that the Hopf map has Hopf invariant 1, and therefore is not null-homotopic. In fact it generates the homotopy group $\pi_3(S^2)$ and has infinite order. In quantum mechanics, the Riemann sphere is known as the Bloch sphere, and the Hopf fibration describes the topological structure of a quantum mechanical two-level system or qubit. Similarly, the topology of a pair of entangled two-level systems is given by the Hopf fibration

$$S^3 \to S^7 \to S^4 \ .$$

(Mosseri & Dandoloff 2001).

Discrete Examples

The regular 4-polytopes: 8-cell (Tesseract), 24-cell, and 120-cell, can each be partitioned into disjoint great circle rings of cells forming discrete Hopf fibrations of these polytopes. The Tesseract partitions into two interlocking rings of four cubes each. The 24-cell partitions into four rings of six octahedrons each. The 120-cell partitions into twelve rings of ten dodecahedrons each.

Classifying Space

In mathematics, specifically in homotopy theory, a classifying space BG of a topological group G is the quotient of a weakly contractible space EG (i.e. a topological space for which all its homotopy groups are trivial) by a free action of G. It has the property that any G principal bundle over a paracompact manifold is isomorphic to a pullback of the principal bundle.

$$EG \to BG.$$

For a discrete group G, BG is, roughly speaking, a path-connected topological space X such that the fundamental group of X is isomorphic to G and the higher homotopy groups of X are trivial, that is, BG is an Eilenberg-Maclane space, or a $K(G,1)$.

Motivation

An example for G infinite cyclic is the circle as X. When G is a discrete group, another way to specify the condition on X is that the universal cover Y of X is contractible. In that case the projection map

$$\pi : Y \rightarrow X$$

becomes a fibre bundle with structure group G, in fact a principal bundle for G. The interest in the classifying space concept really arises from the fact that in this case Y has a universal property with respect to principal G-bundles, in the homotopy category. This is actually more basic than the condition that the higher homotopy groups vanish: the fundamental idea is, given G, to find such a contractible space Y on which G acts *freely*. (The weak equivalence idea of homotopy theory relates the two versions.)

In the case of the circle example, what is being said is that we remark that an infinite cyclic group C acts freely on the real line R, which is contractible. Taking X as the quotient space circle, we can regard the projection π from $R = Y$ to X as a helix in geometrical terms, undergoing projection from three dimensions to the plane. What is being claimed is that π has a universal property amongst principal C-bundles; that any principal C-bundle in a definite way 'comes from'π.

Formalism

A more formal statement takes into account that G may be a topological group (not simply a *discrete group*), and that group actions of G are taken to be continuous; in the absence of continuous actions the classifying space concept can be dealt with, in homotopy terms, via the Eilenberg–MacLane space construction.

In homotopy theory the definition of a topological space BG, the classifying space for principal G-bundles, is given, together with the space EG which is the total space of the universal bundle over BG. That is, what is provided is in fact a continuous mapping

$$\pi : EG \to BG.$$

Assume that the homotopy category of CW complexes is the underlying category, from now on. The *classifying* property required of *BG* in fact relates to π. We must be able to say that given any principal *G*-bundle

$$\gamma : Y \to Z$$

Over a space *Z*, there is a classifying map φ from *Z* to *BG*, such that γ is the pullback of a bundle of π along φ. In less abstract terms, the construction of γ by 'twisting' should be reducible via φ to the twisting already expressed by the construction of π.

For this to be a useful concept, there evidently must be some reason to believe such spaces *BG* exist. In abstract terms (which are not those originally used around 1950 when the idea was first introduced) this is a question of whether the contravariant functor from the homotopy category to the category of sets, defined by $h(Z)$ = set of isomorphism classes of principal *G*-bundles on *Z* is a representable functor. The abstract conditions being known for this (Brown's represen-tability theorem) the result, as an existence theorem, is affirmative and not too difficult.

Applications

This still leaves the question of doing effective calculations with *BG*; for example, the theory of characteristic classes is essentially the same as computing the cohomology groups of *BG*, at least within the restrictive terms of homotopy theory, for interesting groups *G* such as Lie groups. As was shown by the Bott periodicity theorem, the homotopy groups of *BG* are also of fundamental interest. The early work on classifying spaces introduced constructions (for example, the bar construction), that gave concrete descriptions as a simplicial complex.

An example of a classifying space is that when *G* is cyclic of order two; then *BG* is real projective space of infinite dimension, corresponding to the observation that *EG* can be taken as the contractible space resulting from removing the origin in an infinite-dimensional Hilbert space, with *G* acting via *v* going to "*v*, and allowing for homotopy equivalence in choosing BG. This example shows that classifying spaces may be complicated.

In relation with differential geometry (Chern–Weil theory) and the theory of Grassmannians, a much more hands-on approach to the theory is possible for cases such as the unitary groups that are of greatest interest. The construction of the Thom complex MG showed that the spaces BG were also implicated in cobordism theory, so that they assumed a central place in geometric considerations coming out of algebraic topology. Since group cohomology can (in many cases) be defined by the use of classifying spaces, they can also be seen as foundational in much homological algebra.

Generalizations include those for classifying foliations, and the classifying toposes for logical theories of the predicate calculus in intuitionistic logic that take the place of a 'space of models'.

Chapter 7

Borromean Rings

In mathematics, the Borromean rings consist of three topological circles which are linked and form a Brunnian link, i.e., removing any ring results in two unlinked rings.

In other words, no two of the three rings are linked with each other, but nonetheless all three are linked.

Mathematical Properties

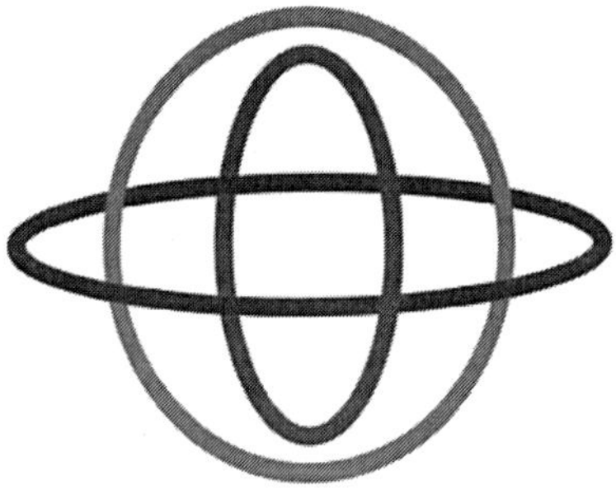

Figure: *A realisation of the Borromean rings as ellipses*

Although the typical picture of the Borromean rings (above right picture) may lead one to think the link can be formed from geometrically round circles, they *cannot* be. (Freedman & Skora 1987) proves that a certain class of links, including the Borromean links, cannot be exactly circular.

Alternatively, this can be seen from considering the link diagram: if one assumes that circles 1 and 2 touch at their two crossing points, then they either lie in a plane or a sphere. In either case, the third circle must pass through this plane or sphere four times, without lying in it, which is impossible.

It is, however, true that one can use ellipses (right picture). These may be taken to be of arbitrarily small eccentricity, i.e. no matter how close to being circular their shape may be, as long as they are not perfectly circular, they can form Borromean links if suitably positioned: for example, Borromean rings made from thin circles of elastic metal wire will bend.

Linking

In knot theory, the Borromean rings are a simple example of a Brunnian link: although each pair of rings are unlinked, the whole link cannot be unlinked. There are a number of ways of seeing this.

Simplest is that the fundamental group of the complement of two unlinked circles is the free group on two generators, a and b, by the Seifert–van Kampen theorem, and then the third loop has the class of the commutator, $[a, b] = aba^{d1}b^{d1}$, as one can see from the link diagram: over one, over the next, back under the first, back under the second.

This is non-trivial in the fundamental group, and thus the Borromean rings are linked. Another way is that the cohomology of the complement supports a non-trivial Massey product, which is not the case for the unlink.

This is a simple example of the Massey product and further, the algebra corresponds to the geometry: a 3-fold Massey product is a 3-fold product which is only defined if all the 2-fold products vanish, which corresponds to the Borromean rings being pairwise unlinked (2-fold products vanish), but linked overall (3-fold product does not vanish). In arithmetic topology, there is an analogy between knots and prime numbers in which one considers links between primes.

The triple of primes (13, 61, 937) are linked modulo 2 (the Rédei symbol is d1) but are pairwise unlinked modulo 2 (the Legendre symbols are all 1). Therefore these primes have been called a "proper Borromean triple modulo 2– or "mod 2 Borromean primes".

Hyperbolic Geometry

The Borromean rings are a hyperbolic link: the complement of the Borromean rings in the 3-sphere admits a complete hyperbolic metric of finite volume. The canonical (Epstein-

Penner) polyhedral decomposition of the complement consists of two regular ideal octahedra. The volume is $16\wedge(\pi/4) = 7.32772\ldots$ where $\wedge$ is the Lobachevsky function.

Connection with Braids

If one cuts the Borromean rings, one obtains one iteration of the standard braid; conversely, if one ties together the ends of (one iteration of) a standard braid, one obtains the Borromean rings. Just as removing one Borromean ring unlinks the remaining two, removing one strand of the standard braid unbraids the other two: they are the basic Brunnian link and Brunnian braid, respectively.

In the standard link diagram, the Borromean rings are ordered non-transitively, in a cyclic order. Using the colours above, these are red over yellow, yellow over blue, blue over red – and thus after removing any one ring, for the remaining two, one is above the other and they can be unlinked. Similarly, in the standard braid, each strand is above one of the others and below the other.

History

The name "Borromean rings" comes from their use in the coat of arms of the aristocratic Borromeo family in Italy. The link itself is much older and has appeared in Gandhara (Afghan) Buddhist art from around the 2nd century, and in the form of the valkn ut on Norse image stones dating back to the 7th century. The Borromean rings have been used in different contexts to indicate strength in unity, e.g., in religion or art. In particular, some have used the design to symbolize the Trinity. The psychoanalyst Jacques Lacan famously found inspiration in the Borromean rings as a model for his topology of human subjectivity, with each ring representing a fundamental Lacanian component of reality (the "real", the "imaginary", and the "symbolic").

The Borromean rings were formerly used as the logo of the German Krupp industrial concern and are used as part of the logo for the successor ThyssenKrupp. The rings were used as the logo of Ballantine beer and are still used by the Ballantine brand beer, now produced by successor Falstaff. In 2006, the International Mathematical Union decided at the 25th

International Congress of Mathematicians in Madrid, Spain to use a new logo based on the Borromean rings. A stone pillar at Marundeeswarar Temple in Thiruvanmiyur, Chennai, Tamil Nadu, India, has such a figure dating to before 6th century.

Partial Rings

In medieval and renaissance Europe, a number of visual signs are found that consist of three elements interlaced together in the same way that the Borromean rings are shown interlaced (in their conventional two-dimensional depiction), but the individual elements are not closed loops. Examples of such symbols are the Snoldelev stone horns and the Diana of Poitiers crescents.

An example with three distinct elements is the logo of Sport Club Internacional. Less-related visual signs include the Gankyil and the Venn diagram on three sets. Similarly, a monkey's fist knot is essentially a 3-dimensional representation of the Borromean rings, albeit with three layers, in most cases. Using the pattern in the incomplete Borromean rings, one can balance three knives on three supports, such as three bottles or glasses, providing a support in the middle for a fourth bottle or glass.

Multiple Rings

Some knot-theoretic links contain multiple Borromean rings configurations; one five-loop link of this type is used as a symbol in Discordianism, based on a depiction in the *Principia Discordia*.

Realisations

Molecular Borromean rings are the molecular counterparts of Borromean rings, which are mechanically-interlocked molecular architectures. In 1997, biologists Chengde Mao and coworkers of New York University succeeded in constructing a set of rings from DNA.

In 2003, chemist Fraser Stoddart and coworkers at UCLA utilised coordination chemistry to construct a set of rings in one step from 18 components. A quantum-mechanical analog of Borromean rings, called an Efimov state, was predicted by physicist Vitaly Efimov in 1970. A team of physicists led by Randall Hulet of Rice University in Houston achieved this with a set of three bound lithium atoms and published their findings in the online journal *Science Express*. In 2010, a team led by K. Tanaka created an Efimov state within a nucleus.

Torus Knot

In knot theory, a torus knot is a special kind of knot that lies on the surface of an unknotted torus in R^3. Similarly, a torus link is a link which lies on the surface of a torus in the same way. Each torus knot is specified by a pair of coprime integers p and q. A torus link arises if p and q are not coprime (in which case the number of components is gcd(p, q). A torus knot is trivial if and only if either p or q is equal to 1 or "1. The simplest nontrivial example is the (2,3)-torus knot, also known as the trefoil knot.

Geometrical Representation

A torus knot can be rendered geometrically in multiple ways which are topologically equivalent but geometrically distinct. The convention used in this article and its figures is the following.

The (p,q)-torus knot winds q times around a circle in the interior of the torus, and p times around its axis of rotational symmetry. If p and q are not relatively prime, then we have a torus link with more than one component. The direction in which the strands of the knot wrap around the torus is also subject to differing conventions. The most common is to have the strands form a right-handed screw for $p\ q > 0$.

The (p,q)-torus knot can be given by the parametrization

$$x = r\cos(p\phi)$$
$$y = r\sin(p\phi)$$
$$z - -\sin(q\phi)$$

where $r = \cos(q\phi) + 2$ and $0 < \phi < 2\pi$. This lies on the surface of the torus given by $(r-2)^2 + z^2 = 1$ (in cylindrical coordinates).

Other parametrizations are also possible, because knots are defined up to continuous deformation. The illustrations for the (2,3)- and (3,8)-torus knots can be obtained by taking $r = \cos(q\phi) + 4$, and in the case of the (2,3)-torus knot by furthermore subtracting respectively $3\cos((p-q)\phi)$ and

$3\sin((p-q)\phi)$ from the above parametrizations of x and y. The latter generalizes smoothly to any coprime p,q satisfying $p<q<2p$

Properties

A torus knot is trivial iff either p or q is equal to 1 or –1. Each torus knot is prime and chiral. The (p,q) torus knot is equivalent to the (q,p) torus knot. This can be proved by moving the strands on the surface of the torus, which is nicely illustrated here.

The (p, q) torus knot is the obverse (mirror image) of the (p,q) torus knot. The $(-p,-q)$ torus knot is equivalent to the (p,q) torus knot except for the reversed orientation.

Any (p,q)-torus knot can be made from a closed braid with p strands. The appropriate braid word is

$$(\sigma_1\sigma_2\cdots\sigma_{p-1})^q.$$

(Please note that this formula assumes the common convention that braid generators are right twists, which is not followed by the Wikipedia page on braids.)

The crossing number of a (p,q) torus knot with $p,q > 0$ is given by

$$c = \min((p-1)q,\ (q-1)p).$$

The genus of a torus knot with $p,q > 0$ is

$$g=\frac{1}{2}(p-1)(q-1).$$

The Alexander polynomial of a torus knot is

$$\frac{(t^{pq}-1)(t-1)}{(t^p-1)(t^q-1)}.$$

The Jones polynomial of a (right-handed) torus knot is given by

$$t^{(p-1)(q-1)/2}\,\frac{1-t^{p+1}-t^{q+1}+t^{p+q}}{1-t^2}.$$

The complement of a torus knot in the 3-sphere is a Seifert-fibered manifold, fibred over the disc with two singular fibres. Let Y be the p-fold dunce cap with a disk removed from the interior,

Z be the q-fold dunce cap with a disk removed its interior, and X be the quotient space obtained by identifying Y and Z along their boundary circle. The knot complement of the (p, q)-torus knot deformation retracts to the space X. Therefore, the knot group of a torus knot has the presentation

$$\langle x, y \mid x^p = y^q \rangle.$$

Torus knots are the only knots whose knot groups have nontrivial centre (which is infinite cyclic, generated by the element $x^p = y^q$ in the presentation above).

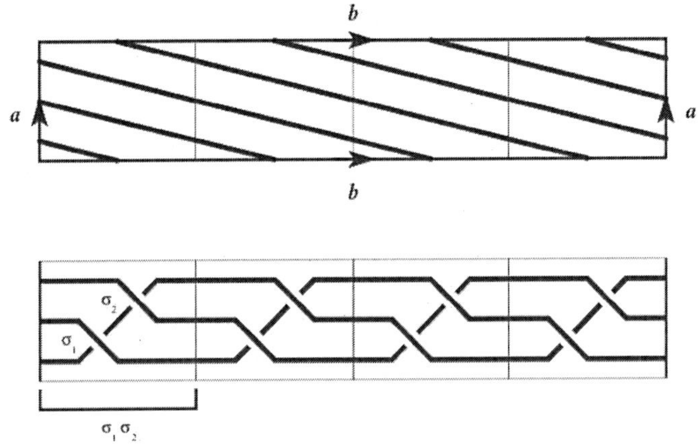

Figure: *The (3, 4) torus knot on the unwrapped torus surface, and its braid word*

Prime Knot

In knot theory, a prime knot is a knot that is, in a certain sense, indecomposable. Specifically, it is a non-trivial knot which cannot be written as the knot sum of two non-trivial knots. Knots that are not prime are said to be composite. It can be a nontrivial problem to determine whether a given knot is prime or not. A nice family of examples of prime knots are the torus knots.

These are formed by wrapping a circle around a torus p times in one direction and q times in the other, where p and q are coprime integers.

The simplest prime knot is the trefoil with three crossings. The trefoil is actually a (2, 3)-torus knot. The figure-eight knot, with four crossings, is the simplest non-torus knot. For any positive integer n, there are a finite number of prime knots with n crossings. The first few values (sequence A002863 in OEIS) are given in the following table.

n	1	2	3	4	5	6	7	8	9	10	11	12	13	14	15	16
Number of prime knots with n crossings	0	0	1	1	2	3	7	21	49	165	552	2176	9988	46972	253293	1388705

Note that enantiomorphs are counted only once in this table and the following chart (i.e. a knot and its mirror image are considered equivalent).

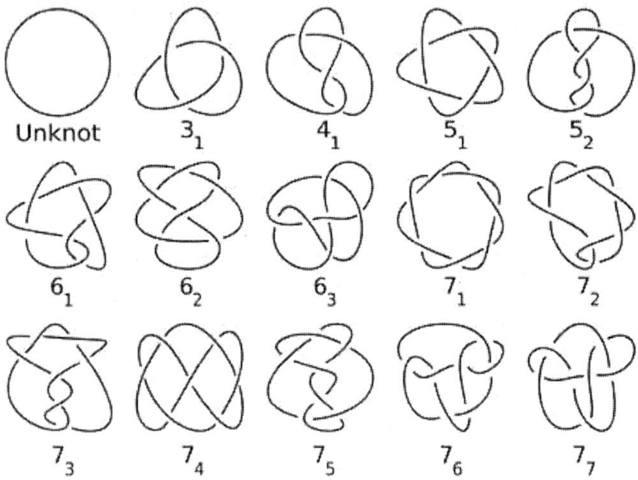

Figure: *A chart of all prime knots with seven or fewer crossings, not including mirror-images. (The unknot is not considered prime.)*

Schubert's Theorem

A theorem due to Horst Schubert states that every knot can be uniquely expressed as a connected sum of prime knots.

Alternating Knot

In knot theory, a link diagram is alternating if the crossings alternate under, over, under, over, as you travel along each component of the link. A link is alternating if it has an alternating diagram. Many of the knots with crossing number less than 10 are alternating. This fact and useful properties of alternating knots, such as the Tait conjectures, was what enabled early knot tabulators, such as Tait, to construct tables with relatively few mistakes or omissions.

The simplest non-alternating prime knots have 8 crossings (and there are three such). It is conjectured that as the crossing number increases, the percentage of knots that are alternating goes to 0 exponentially quickly.

Alternating links end up having an important role in knot theory and 3-manifold theory, due to their complements having useful and interesting geometric and topological properties. This led Ralph Fox to ask, "What is an alternating knot?" By this he was asking what non-diagrammatic properties of the knot complement would characterise alternating knots.

Various geometric and topological information is revealed in an alternating diagram. Primeness and splittability of a link is easily seen from the diagram. The crossing number of a reduced, alternating diagram is the crossing number of the knot.

This last is one of the celebrated Tait Conjectures. An alternating knot diagram is in one to one correspondence with a planar graph. Each crossing is associated with an edge and half of the connected components of the complement of the diagram are associated with vertices in a checker board manner.

Tait Conjectures

The Tait conjectures are:

1. Any reduced diagram of an alternating link has the fewest possible crossings.
2. Any two reduced diagrams of the same alternating knot have the same writhe.
3. Given any two reduced alternating diagrams D_1 and D_2 of an oriented, prime alternating link: D_1 may be transformed to D_2 by means of a sequence of certain simple moves called *flypes*. Also known as the Tait flyping conjecture.

Morwen Thistlethwaite, Louis Kauffman and K. Murasugi proved the first two Tait conjectures in 1987 and Morwen Thistlethwaite and William Menasco proved the Tait flyping conjecture in 1991.

Hyperbolic Volume

Menasco, applying Thurston's hyperbolization theorem for Haken manifolds, showed that any prime, non-split alternating link is hyperbolic, i.e. the link complement has a hyperbolic geometry, unless the link is a torus link.

Thus hyperbolic volume is an invariant of many alternating links. Marc Lackenby has shown that the volume has upper and lower linear bounds as functions of the number of *twist regions* of a reduced, alternating diagram.

Hyperbolic Link

In mathematics, a hyperbolic link is a link in the 3-sphere with complement that has a complete Riemannian metric of constant negative curvature, i.e. has a hyperbolic geometry. A hyperbolic knot is a hyperbolic link with one component.

As a consequence of the work of William Thurston, it is known that every knot is precisely one of the following: hyperbolic, a torus knot, or a satellite knot. As a consequence, hyperbolic knots can be considered plentiful. A similar heuristic applies to hyperbolic links.

As a consequence of Thurston's hyperbolic Dehn surgery theorem, performing Dehn surgeries on a hyperbolic link enables one to obtain many more hyperbolic 3-manifolds.

Examples

- Borromean rings are hyperbolic.
- Every non-split, prime, alternating link that is not a torus link is hyperbolic by a result of William Menasco.

Crossing Number (Knot Theory)

In the mathematical area of knot theory, the crossing number of a knot is the minimal number of crossings of any diagram of the knot. It is a knot invariant.

By way of example, the unknot has crossing number zero, the trefoil knot three and the figure-eight knot four. There are no other knots with a crossing number this low, and just two knots have crossing number five, but the number of knots with a particular crossing number increases rapidly as the crossing number increases.

Tables of prime knots are traditionally indexed by crossing number, with a subscript to indicate which particular knot out of those with this many crossings is meant (this sub-ordering is not based on anything in particular, except that torus knots are listed first). The listing goes 3_1 (the trefoil knot), 4_1 (the figure-eight knot), 5_1, 5_2, 6_1, etc. This order has not changed significantly since P. G. Tait published a tabulation of knots in 1877.

There has been very little progress on understanding the behaviour of crossing number under rudimentary operations on knots. A big open question asks if the crossing number is additive when taking knot sums. It is also expected that a satellite of a knot K should have larger crossing number than K, but this has not been proven.

Additivity of crossing number under knot sum has been proven for special cases, for example if the summands are alternating knots (or more generally, adequate knot), or if the summands are torus knots.

Marc Lackenby has also given a proof that there is a constant $N > 1$ such that $\frac{1}{N}(\mathrm{cr}(K_1)+\mathrm{cr}(K_2)) \leq \mathrm{cr}(K_1 + K_2)$, but his method, which utilizes normal surfaces, cannot improve N to 1. There are related concepts of average crossing number and asymptotic crossing number. Both of these quantities bound the standard crossing number. Asymptotic crossing number is conjectured to be equal to crossing number.

There are mysterious connections between the crossing number of a knot and the physical behaviour of DNA knots. For prime DNA knots, crossing number is a good predictor of the relative velocity of the DNA knot in agarose gel electrophoresis. Basically, the higher the crossing number, the faster the relative velocity. For composite knots, this does not appear to be the case, although experimental conditions can drastically change the results.

Knot Invariant

In the mathematical field of knot theory, a knot invariant is a quantity (in a broad sense) defined for each knot which is the same for equivalent knots. The equivalence is often given by ambient isotopy but can be given by homeomorphism. Some invariants are indeed numbers, but invariants can range from the simple, such as a yes/no answer, to those as complex as a homology theory.

Research on invariants is not only motivated by the basic problem of distinguishing one knot from another but also to understand fundamental properties of knots and their relations to other branches of mathematics. From the modern perspective, it is natural to define a knot invariant from a knot diagram.

Of course, it must be unchanged (that is to say, invariant) under the Reidemeister moves. Tricolorability is a particularly simple example.

Other examples are knot polynomials, such as the Jones polynomial, which are currently among the most useful invariants for distinguishing knots from one another, though at the time of writing it is not known whether there exists a knot polynomial which distinguishes all knots from each other, or even which distinguishes just the unknot from all other knots.

Other invariants can be defined by considering some integer-valued function of knot diagrams and taking its minimum value over all possible diagrams of a given knot. This category includes the crossing number, which is the minimum number of crossings for any diagram of the knot, and the bridge number, which is the minimum number of bridges for any diagram of the knot.

Historically, many of the early knot invariants are not defined by first selecting a diagram but defined intrinsically, which can make computing some of these invariants a challenge. For example, knot genus is particularly tricky to compute, but can be effective (for instance, in distinguishing mutants).

The complement of a knot itself (as a topological space) is known to be a "complete invariant" of the knot by the Gordon–Luecke theorem in the sense that it distinguishes the given knot from all other knots up to ambient isotopy and mirror image. Some invariants associated with the knot complement include the knot group which is just the fundamental group of the complement.

The knot quandle is also a complete invariant in this sense but it is difficult to determine if two quandles are isomorphic.

By Mostow–Prasad rigidity, the hyperbolic structure on the complement of a hyperbolic link is unique, which means the hyperbolic volume is an invariant for these knots and links. Volume, and other hyperbolic invariants, have proven very effective, utilized in some of the extensive efforts at knot tabulation.

In recent years, there has been much interest in homological invariants of knots which categorify well-known invariants. Heegaard Floer homology is a homology theory whose Euler characteristic is the Alexander polynomial of the knot. It has been proven effective in deducing new results about the classical invariants.

Along a different line of study, there is a combinatorially defined cohomology theory of knots called Khovanov homology whose Euler characteristic is the Jones polynomial. This has recently been shown to be useful in obtaining bounds on slice genus whose earlier proofs required gauge theory. Khovanov and Rozansky have since defined several other related cohomology theories whose Euler characteristics recover other classical invariants.

Stroppel gave a representation theoretic interpretation of Khovanov homology by categorifying quantum group invariants. There is also growing interest from both knot theorists and scientists in understanding "physical" or geometric properties of knots and relating it to topological invariants and knot type. An old result in this direction is the Fary–Milnor theorem states that if the total curvature of a knot K in $\mathbb{R}^3$ satisfies

$$\oint_K \kappa ds < 4\pi,$$

where $\kappa(p)$ is the curvature at p, then K is an unknot. Therefore, for knotted curves,

$$\oint_K \kappa ds > 4\pi.$$

An example of a "physical" invariant is ropelength, which is the amount of 1-inch diametre rope needed to realise a particular knot type.

Linking Number

In mathematics, the linking number is a numerical invariant that describes the linking of two closed curves in three-dimensional space. Intuitively, the linking number represents the number of times that each curve winds around the other.

The linking number is always an integer, but may be positive or negative depending on the orientation of the two curves.

The linking number was introduced by Gauss in the form of the linking integral.

It is an important object of study in knot theory, algebraic topology, and differential geometry, and has numerous applications in mathematics and science, including quantum mechanics, electro-magnetism, and the study of DNA supercoiling.

Computing the Linking Number

There is an algorithm to compute the linking number of two curves from a link diagram. Label each crossing as *positive* or *negative*, acco-rding to the following rule:

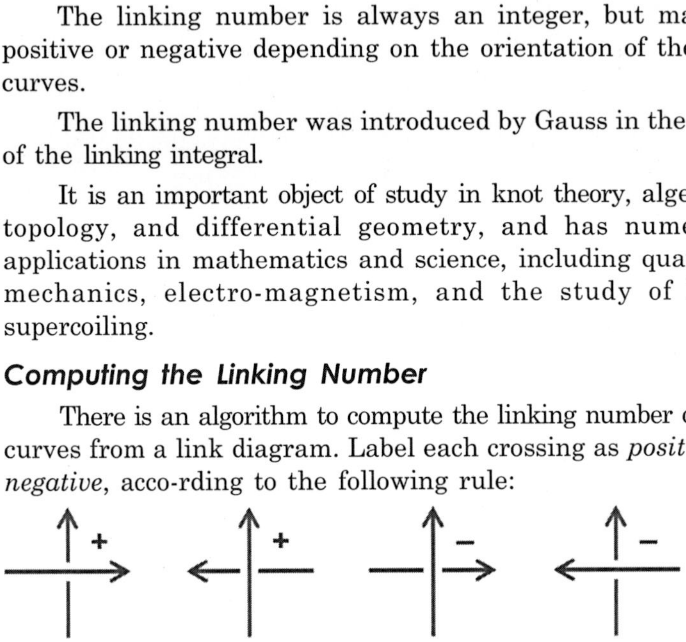

The total number of positive crossings minus the total number of negative crossings is equal to *twice* the linking number. That is:

$$\text{linking number} = \frac{n_1 + n_2 - n_3 - n_4}{2}$$

where n_1, n_2, n_3, n_4 represent the number of crossings of each of the four types. The two sums $n_1 + n_3$ and $n_2 + n_4$ are always equal, which leads to the following alternative formula

$$\text{linking number} = n_1 - n_4 = n_2 - n_3.$$

Note that $n_1 - n_4$ involves only the undercrossings of the blue curve by the red, while $n_2 - n_3$ involves only the overcrossings.

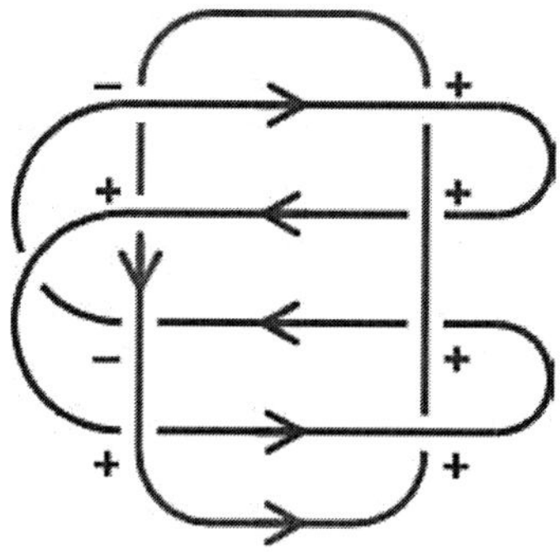

Figure: *With six positive crossings and two negative crossings, these curves have linking number two.*

Properties and Examples

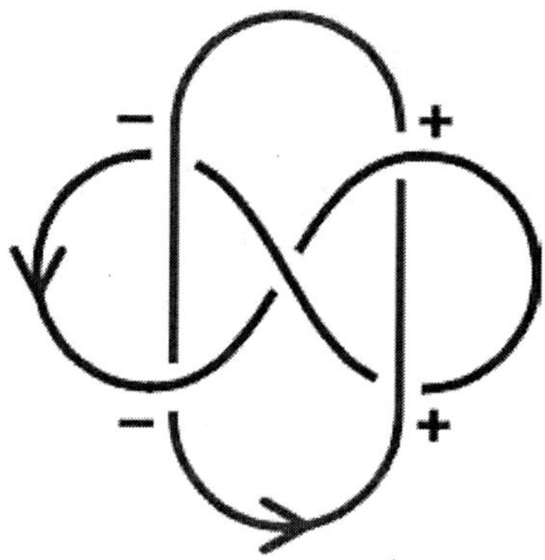

Figure: *The two curves of the Whitehead link have linking number zero.*

- Any two unlinked curves have linking number zero. However, two curves with linking number zero may still be linked (e.g. the Whitehead link).
- Reversing the orientation of either of the curves negates the linking number, while reversing the orientation of both curves leaves it unchanged.
- The linking number is chiral: taking the mirror image of link negates the linking number. Our convention for positive linking number is based on a right-hand rule.
- The winding number of an oriented curve in the x-y plane is equal to its linking number with the z-axis (thinking of the z-axis as a closed curve in the 3-sphere).

- More generally, if either of the curves is simple, then the first homology group of its complement is isomorphic to Z. In this case, the linking number is determined by the homology class of the other curve.
- In physics, the linking number is an example of a topological quantum number. It is related to quantum entanglement.

Gauss's Integral Definition

Given two non-intersecting differentiable curves $\gamma_1, \gamma_2 : S^1 \to \mathbb{R}^3$, define the Gauss map Γ from the torus to the sphere by

$$\Gamma(s,t) = \frac{\gamma_1(s) - \gamma_2(t)}{|\gamma_1(s) - \gamma_2(t)|}.$$

Pick a point in the unit sphere, v, so that orthogonal projection of the link to the plane perpendicular to v gives a link diagram. Observe that a point *(s,t)* that goes to v under the Gauss map corresponds to a crossing in the link diagram where γ_1 is over γ_2. Also, a neighbourhood of *(s,t)* is mapped under the Gauss map to a neighbourhood of v preserving or reversing orientation depending on the sign of the crossing.

Thus in order to compute the linking number of the diagram corresponding to v it suffices to count the *signed* number of times the Gauss map covers v. Since v is a regular value, this is precisely the degree of the Gauss map (i.e. the signed number of times that the image of Γ covers the sphere). Isotopy invariance of the linking number is automatically obtained as the degree is invariant under homotopic maps. Any other regular value would give the same number, so the linking number doesn't depend on any particular link diagram.

This formulation of the linking number of γ_1 and γ_2 enables an explicit formula as a double line integral, the Gauss linking integral:

$$\text{linking number} = \frac{1}{4\pi} \oint_{\gamma_1} \oint_{\gamma_2} \frac{\mathbf{r}_1 - \mathbf{r}_2}{|\mathbf{r}_1 - \mathbf{r}_2|^3} \cdot (d\mathbf{r}_1 \times d\mathbf{r}_2).$$

This integral computes the total signed area of the image of the Gauss map (the integrand being the Jacobian of Γ) and then divides by the area of the sphere (which is 4π).

Generalizations

- Just as closed curves can be linked in three dimensions, any two closed manifolds of dimensions m and n may be linked in a Euclidean space of dimension $m+n+1$. Any such link has an associated Gauss map, whose degree is a generalization of the linking number.
- Any framed knot has a self-linking number obtained by computing the linking number of the knot C with a new curve obtained by slightly moving the points of C along the framing vectors. The self-linking number obtained by moving vertically (along the blackboard framing) is known as Kauffman's self-linking number.
- The linking number is defined for two linked circles; given three or more circles, one can define the Milnor invariants, which are a numerical invariant generalizing linking number.
- In algebraic topology, the cup product is a far-reaching algebraic generalization of the linking number, with the Massey products being the algebraic analogs for the Milnor invariants.
- A linkless embedding of an undirected graph is an embedding into three-dimensional space such that every two cycles have zero linking number. The graphs that have a linkless embedding have a forbidden minor characterisation as the graphs with no Petersen family minor.

Skein Relation

A central question in the mathematical theory of knots is whether two knot diagrams represent the same knot. One tool used to answer such questions is a knot polynomial which is an invariant of the knot. If two diagrams have different polynomials, they represent different knots. The reverse may not be true. Skein relations are often used to give a simple definition of knot polynomials. Informally, a skein relation gives a linear relation between the values of a knot polynomial on a collection of three links which differ from each other only in a small region. For some knot polynomials, such as the Conway, Alexander, and Jones polynomials, the relevant skein relations are sufficient to calculate the polynomial recursively. For others, such as the HOMFLYPT polynomial, more complicated algorithms are necessary.

Definition

A skein relationship requires three link diagrams that are identical except at one crossing. The three diagrams must exhibit the three possibilities that could occur at that crossing, it could be *under,* it could be *over* or it could not exist at all. Link diagrams must be considered because a single skein change can alter a diagram from representing a knot to representing a link and vice versa. Depending on the knot polynomial in question, the links (or tangles) appearing in a skein relation may be oriented or unoriented.

The three diagrams are labelled as follows. Turn the diagrams so the directions at the crossing in question are both roughly northward. One diagram will have northwest over northeast, it is labelled L_-. Another will have northeast over northwest, it's L_+. The remaining diagram is lacking that crossing and is labelled L_0.

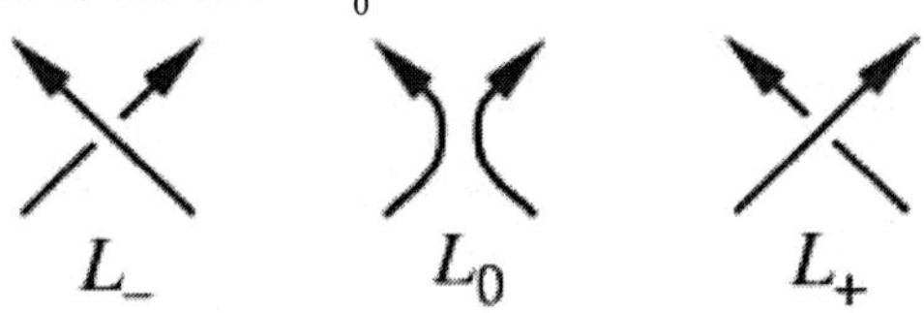

(The labelling is actually independent of direction insofar as it remains the same if all directions are reversed. Thus polynomials on undirected knots are unambiguously defined by this method. However, the directions on *links* are a vital detail to retain as one recurses through a polynomial calculation.)

It is also sensible to think in a generative sense, by taking an existing link diagram and "patching" it to make the other two—just so long as the patches are applied with compatible directions.

To recursively define a knot (link) polynomial, a function F is fixed and for any triple of diagrams and their polynomials labelled as above,

$$F(L_-, L_0, L_+) = 0$$

or more pedantically

$$F\left(L_{-}(x), L_{0}(x), L_{+}(x), x\right) = 0 \text{ for all } x$$

(Finding an F which produces polynomials independent of the sequences of crossings used in a recursion is no trivial exercise.)

More formally, a skein relation can be thought of as defining the kernel of a quotient map from the planar algebra of tangles. Such a map corresponds to a knot polynomial if all closed diagrams are taken to some (polynomial) multiple of the image of the empty diagram.

Knot Polynomial

In the mathematical field of knot theory, a knot polynomial is a knot invariant in the form of a polynomial whose coefficients encode some of the properties of a given kn.

History

The first knot polynomial, the Alexander polynomial, was introduced by James Waddell Alexander II in 1923, but other knot polynomials were not found until almost 60 years later.

In the 1960s, John Conway came up with a skein relation for a version of the Alexander polynomial, usually referred to as the Alexander–Conway polynomial. The significance of this skein relation was not realised until the early 1980s, when Vaughan Jones discovered the Jones polynomial. This led to the discovery of more knot polynomials, such as the so-called HOMFLY polynomial.

Soon after Jones' discovery, Louis Kauffman noticed the Jones polynomial could be computed by means of a state-sum model, which involved the bracket polynomial, an invariant of framed knots. This opened up avenues of research linking knot theory and statistical mechanics.

In the late 1980s, two related breakthroughs were made. Edward Witten demonstrated that the Jones polynomial, and similar Jones-type invariants, had an interpretation in Chern–Simons theory. Viktor Vassiliev and Mikhail Goussarov started the theory of finite type invariants of knots.

The coefficients of the previously named polynomials are known to be of finite type (after perhaps a suitable "change of variables").

In recent years, the Alexander polynomial has been shown to be related to Floer homology. The graded Euler characteristic of the knot Floer homology of Ozsvαth and Szabσ is the Alexander polynomial.

Alexander Polynomial

In mathematics, the Alexander polynomial is a knot invariant which assigns a polynomial with integer coefficients to each knot type. James Waddell Alexander II discovered this, the first knot polynomial, in 1923. In 1969, John Conway showed a version of this polynomial, now called the Alexander–Conway polynomial, could be computed using a skein relation, although its significance was not realised until the discovery of the Jones polynomial in 1984.

Soon after Conway's reworking of the Alexander polynomial, it was realised that a similar skein relation was exhibited in Alexander's paper on his polynomial.

Definition

Let K be a knot in the 3-sphere. Let X be the infinite cyclic cover of the knot complement of K. This covering can be obtained by cutting the knot complement along a Seifert surface of K and gluing together infinitely many copies of the resulting manifold with boundary in a cyclic manner. There is a covering transformation t acting on X. Consider the first homology (with integer coefficients) of X, denoted $H1(X)$. The transformation t acts on the homology and so we can consider a module over. This is called the Alexander invariant or Alexander module.

The module is finitely presentable; a presentation matrix for this module is called the Alexander matrix. If the number of generators, r, is less than or equal to the number of relations, s, then we consider the ideal generated by all r by r minors of the matrix; this is the zero'th Fitting ideal or Alexander ideal and does not depend on choice of presentation matrix.

If $r > s$, set the ideal equal to 0. If the Alexander ideal is principal, take a generator; this is called an Alexander polynomial of the knot. Since this is only unique up to multiplication by the Laurent monomial , one often fixes a particular unique form. Alexander's choice of normalization is to make the polynomial have a positive constant term.

Alexander proved that the Alexander ideal is nonzero and always principal. Thus an Alexander polynomial always exists, and is clearly a knot invariant, denoted.

Computing the Polynomial

The following procedure for computing the Alexander polynomial was given by J. W. Alexander in his paper. Take an oriented diagram of the knot with n crossings; there are $n + 2$ regions of the knot diagram. To work out the Alexander polynomial, first one must create an incidence matrix of size $(n, n + 2)$. The n rows correspond to the n crossings, and the $n + 2$ columns to the regions. The values for the matrix entries are either 0, 1, d1, t, dt.

Consider the entry corresponding to a particular region and crossing. If the region is not adjacent to the crossing, the entry is 0. If the region is adjacent to the crossing, the entry depends on its location. The following table gives the entry, determined by the location of the region at the crossing from the perspective of the incoming undercrossing line.

on the left before undercrossing: $-t$

on the right before undercrossing: 1

on the left after undercrossing: t

on the right after undercrossing: -1

Remove two columns corresponding to adjacent regions from the matrix, and work out the determinant of the new n by n matrix. Depending on the columns removed, the answer will differ by multiplication by $\pm t^n$. To resolve this ambiguity, divide out the largest possible power of t and multiply by d1 if necessary, so that the constant term is positive. This gives the Alexander polynomial. The Alexander polynomial can also be computed from the Seifert matrix.

Geometric Significance of the Polynomial

Since the Alexander ideal is principal, $\Delta_K(t) = 1$ if and only if the commutator subgroup of the knot group is perfect (i.e. equal to its own commutator subgroup). For a topologically slice knot, the Alexander polynomial satisfies the Fox–Milnor condition $\Delta_K(t) = f(t)f(t^{-1})$ where $f(t)$ is some other integral Laurent polynomial.

Twice the knot genus is bounded below by the degree of the Alexander polynomial.

Michael Freedman proved that a knot in the 3-sphere is topologically slice; i.e., bounds a "locally-flat" topological disc in the 4-ball, if the Alexander polynomial of the knot is trivial (Freedman and Quinn, 1990).

There are other relations with surfaces and smooth 4-dimensional topology. For example, under certain assumptions, there is a way of modifying a smooth 4-manifold by performing a surgery that consists of removing a neighbourhood of a two-dimensional torus and replacing it with a knot complement crossed with S^1. The result is a smooth 4-manifold homeomorphic to the original, though now the Seiberg–Witten invariant has been modified by multiplication with the Alexander polynomial of the knot.

Knots with symmetries are known to have restricted Alexander polynomials. Although, the Alexander polynomial can fail to detect some symmetries, such as strong invertibility. If the knot complement fibres over the circle, then the Alexander polynomial of the knot is known to be *monic* (highest and lowest order terms equal to ± 1). In fact, if $S \to C_K \to S^1$ is a fibre bundle where C_K is the knot complement, let $g : S \to S$ represent the monodromy, then $\Delta_K(t) = Det(tI - g_*)$ where $g_* : H_1 S \to H_1 S$ is the induced map on homology.

Alexander–Conway Polynomial

Alexander proved the Alexander polynomial satisfies a skein relation. John Conway later rediscovered this in a different form and showed that the skein relation together with a choice of value on the unknot was enough to determine the polynomial. Conway's version is a polynomial in z with integer coefficients, denoted $\nabla(z)$ and called the Alexander–Conway polynomial (also known as Conway polynomial or Conway–Alexander polynomial).

Suppose we are given an oriented link diagram, where L_+, L_-, L_0 are link diagrams resulting crossing and smoothing changes on a local region of a specified crossing of the diagram, as indicated in the figure.

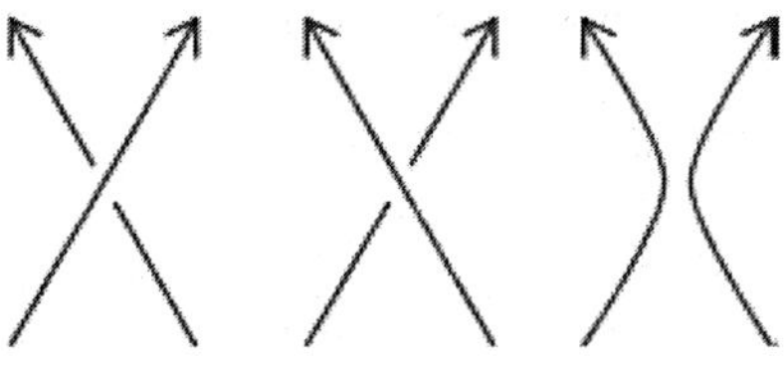

Here are Conway's skein relations:

- $\nabla(O) = 1$ (where O is any diagram of the unknot)
- $\nabla(L_+) - \nabla(L_-) = z\nabla(L_0)$

The relationship to the standard Alexander polynomial is given by $\Delta_L(t^2) = \nabla_L(t - t^{-1})$. Here Δ_L must be properly normalized (by multiplication of $\pm t^{n/2}$) to satisfy the skein relation $\Delta(L_+) - \Delta(L_-) = (t^{1/2} - t^{-1/2})\Delta(L_0)$ Note that this relation gives a Laurent polynomial in $t^{1/2}$.

Knot Group

In mathematics, a knot is an embedding of a circle into 3-dimensional Euclidean space. The knot group of a knot K is defined as the fundamental group of the knot complement of K in R^3,

$$\pi_1(\mathbb{R}^3 \setminus K).$$

Other conventions consider knots to be embedded in the 3-sphere, in which case the knot group is the fundamental group of its complement in S^3.

Properties

Two equivalent knots have isomorphic knot groups, so the knot group is a knot invariant and can be used to distinguish between inequivalent knots. This is because an equivalence between two knots is a self homeomorphism of $\mathbb{R}^3$ isotopic to the identity sending the first knot onto the second.

Clearly the homeomorphism restricts onto a homeomorphism of the complement of the knot therefore inducing an isomorphism in the fundamental group of the knot complement. However, two knots can have isomorphic knot groups without necessarily being equivalent.

The abelianization of a knot group is always isomorphic to the infinite cyclic group Z; this follows because the abelianization agrees with the first homology group, which can be easily computed.

The knot group (or fundamental group of an oriented link in general) can be computed in the Wirtinger presentation by a relatively simple algorithm.

Examples

- The unknot has knot group isomorphic to Z.
- The trefoil knot has knot group isomorphic to the braid group B_3. This group has the presentation

  $$\langle x, y \mid x^2 = y^3 \rangle \text{or} \langle a, b \mid aba = bab \rangle.$$

- A (p,q)-torus knot has knot group with presentation

  $$\langle x, y \mid x^p = y^q \rangle$$

- The figure eight knot has knot group with presentation

  $$\langle x, y \mid x^{-1}yxy^{-1}xy = yx^{-1}yx \rangle$$

- The square knot and the granny knot have isomorphic knot groups, yet these two knots are inequivalent.

Writhe

In knot theory, the writhe is a property of an oriented link diagram. The writhe is the total number of *positive crossings* minus the total number of *negative crossings*.

A direction is assigned to the link at a point in each component and this direction is followed all the way around each component. If as you travel along a link component and cross over a crossing, the strand underneath goes from right to left, the crossing is positive; if the lower strand goes from left to right, the crossing is negative. One way of remembering this is to use a variation of the right-hand rule.

For a *knot* diagram, using the right-hand rule with either orientation gives the same result, so the writhe is well-defined on unoriented knot diagrams.

The writhe of a knot is unaffected by two of the three Reidemeister moves: moves of Type II and Type III do not affect the writhe. Reidemeister move Type I, however, increases or decreases the writhe by 1. This implies that the writhe of a knot is *not* an isotopy invariant of the knot itself — only the diagram. By a series of Type I moves one can set the writhe of a diagram for a given knot to be any integer at all.

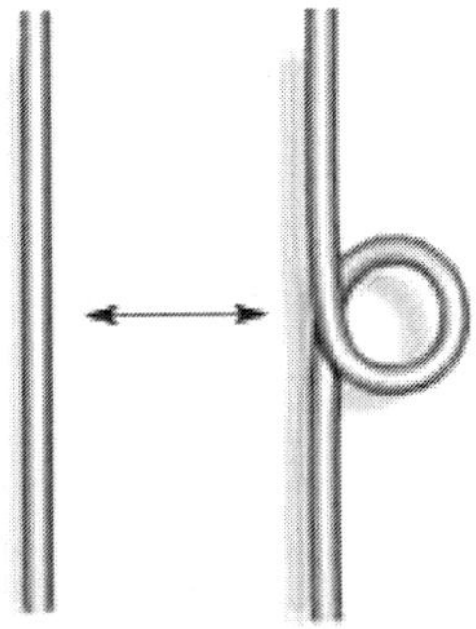

Figure: *A Type I Reidemeister move changes the writhe by 1*

Racks and Quandles

In mathematics, racks and quandles are sets with a binary operations satisfying axioms analogous to the Reidemeister moves used to manipulate knot diagrams. While mainly used to obtain invariants of knots, they can be viewed as algebraic constructions in their own right. In particular, the definition of a quandle axiomatizes the properties of conjugation in a group.

History

The earliest known work on racks is contained within (unpublished) 1959 correspondence between John Conway and Gavin Wraith, who at the time were undergraduate students at the University of Cambridge. Wraith had become interested in these structures (which he initially dubbed sequentials) while at school. Conway renamed them wracks, partly as a pun on his colleague's name, and partly because they arise as the remnants (or 'wrack and ruin') of a group when one discards the multiplicative structure and considers only the conjugation structure. The spelling 'rack' has now become prevalent.

These constructs surfaced again in the 1980s: in a 1982 paper by David Joyce (where the term quandle was coined), in a 1982 paper by Sergei Matveev (under the name distributive groupoids) and in a 1986 conference paper by Egbert Brieskorn (where they were called automorphic sets). A detailed overview of racks and their applications in knot theory may be found in the paper by Colin Rourke and Roger Fenn.

Quandles

A quandle is defined as a rack Q such that $\forall a \in Q$

$$a \triangleright a = a$$

or equivalently

$$a \triangleleft a = a$$

Examples and Applications

Every group gives a quandle where the operations come from conjugation:

$$a \triangleright b = aba^{-1}$$

$$b \triangleleft a = a^{-1}ba$$

In fact, every equational law satisfied by conjugation in a group follows from the quandle axioms. So, one can think of a quandle as what is left of a group when we forget multiplication, the identity, and inverses, and only remember the operation of conjugation.

Every tame knot in three dimensional Euclidean space has a 'fundamental quandle'. To define this, one can note that the fundamental group of the knot complement, or knot group, has a presentation (the Wirtinger presentation) in which the relations only involve conjugation. So, this presentation can also be used as a presentation of a quandle. The fundamental quandle is a very powerful invariant of knots. In particular, if two knots have isomorphic fundamental quandles then there is a homeomorphism of three-dimensional Euclidean space, which may be orientation reversing, taking one knot to the other.

Less powerful but more easily computable invariants of knots may be obtained by counting the homomorphisms from the knot quandle to a fixed quandle Q. Since the Wirtinger presentation has one generator for each strand in a knot diagram, these invariants can be computed by counting ways of labelling each strand by an element of , subject to certain constraints. More sophisticated invariants of this sort can be constructed with the help of quandle cohomology.

The Alexander quandles are also important, since they can be used to compute the Alexander polynomial of a knot. Let be a module over the ring $\mathbb{Z}[t,t^{-1}]$ of Laurent polynomials in one variable. Then the Alexander quandle is made into a quandle with the left action given by

$$a \triangleright b = ta + (1-t)b$$

Racks are a useful generalization of quandles in topology, since while quandles can represent knots on a round linear object (such as rope or a thread), racks can represent ribbons, which may be twisted as well as knotted.

A quandle Q is said to be involutory if for all $a, b \in Q$

$$a \triangleright (a \triangleright b) = b$$

or equivalently

$$(b \triangleleft a) \triangleleft a = b$$

Any symmetric space gives an involutory quandle, where $a \triangleright b$ is the result of 'reflecting b through a'.

Seifert Surface

In mathematics, a Seifert surface (named after German mathematician Herbert Seifert) is a surface whose boundary is a given knot or link. Such surfaces can be used to study the properties of the associated knot or link. For example, many knot invariants are most easily calculated using a Seifert surface. Seifert surfaces are also interesting in their own right, and the subject of considerable research.

Specifically, let L be a tame oriented knot or link in Euclidean 3-space (or in the 3-sphere). A Seifert surface is a compact, connected, oriented surface S embedded in 3-space whose boundary is L such that the orientation on L is just the induced orientation from S, and every connected component of S has non-empty boundary.

Note that any compact, connected, oriented surface with nonempty boundary in Euclidean 3-space is the Seifert surface associated to its boundary link. A single knot or link can have many different inequivalent Seifert surfaces. A Seifert surface must be oriented. It is possible to associate unoriented (and not necessarily orientable) surfaces to knots as well.

Examples

The standard Möbius strip has the unknot for a boundary but is not considered to be a Seifert surface for the unknot because it is not orientable.

The "checkerboard" coloring of the usual minimal crossing projection of the trefoil knot gives a Mobius strip with three half twists. As with the previous example, this is not a Seifert surface as it is not orientable. Applying Seifert's algorithm to this diagram, as expected, does produce a Seifert surface; in this case, it is a punctured torus of genus $g=1$, and the Seifert matrix is

$$V = \begin{pmatrix} 1 & -1 \\ 0 & 1 \end{pmatrix}.$$

Existence and Seifert Matrix

It is a theorem that any link always has an associated Seifert surface. This theorem was first published by Frankl and Pontrjagin in 1930. A different proof was published in 1934 by Herbert Seifert and relies on what is now called the Seifert algorithm. The algorithm produces a Seifert surface S, given a projection of the knot or link in question.

Suppose that link has m components (m=1 for a knot), the diagram has d crossing points, and resolving the crossings (preserving the orientation of the knot) yields f circles. Then the surface is constructed from f disjoint disks by attaching d bands. The homology group is free abelian on $2g$ generators, where

$$g = (2 + d - f - m)/2$$

is the genus of the intersection form Q on $H_1(S)$ is skew-symmetric, and there is a basis of $2g$ cycles

$$a_1, a_2, \ldots, a_{2g}$$

with

$$Q=(Q(a_i, a_j))$$

the direct sum of g copies of

$$\begin{pmatrix} 0 & -1 \\ 1 & 0 \end{pmatrix}.$$

The $2g\times 2g$ integer Seifert matrix

$$V=(v(i,j))$$

has $v(i,j)$ the linking number in Euclidean 3-space (or in the 3-sphere) of a_i and the pushoff of a_j out of the surface, with

$$V-V^{*}=Q$$

where $*V=(v(j,i))$ the transpose matrix. Every integer $2g\times 2g$ matrix V with $V-V^{*}=Q$ arises as the Seifert matrix of a knot with genus g Seifert surface.

The Alexander polynomial is computed from the Seifert matrix by $A(t)=det(V-tV\,,)$ which is a polynomial in the indeterminate of degree $\leq 2g$.

The Alexander polynomial is independent of the choice of Seifert surface , and is an invariant of the knot or link.

The signature of a knot is the signature of the symmetric Seifert matrix $V+V^{\top}$. It is again an invariant of the knot or link.

Genus of a Knot

Seifert surfaces are not at all unique: a Seifert surface S of genus g and Seifert matrix V can be modified by a surgery, to be replaced by a Seifert surface S' of genus $g+1$ and Seifert matrix

$$V'=V\oplus\begin{pmatrix}0 & 1\\ 1 & 0\end{pmatrix}.$$

The genus of a knot K is the knot invariant defined by the minimal genus g of a Seifert surface for K.

For instance:

- An unknot—which is, by definition, the boundary of a disc—has genus zero. Moreover, the unknot is the *only* knot with genus zero.
- The trefoil knot has genus one, as does the figure-eight knot.
- The genus of a (p,q)-torus knot is $(p-1)(q-1)/2$
- The degree of the Alexander polynomial is a lower bound on twice the genus of the knot.

A fundamental property of the genus is that it is additive with respect to the knot sum:

$$g(K_1 \# K_2) = g(K_1) + g(K_2)$$

Braid Theory

In topology, a branch of mathematics, braid theory is an abstract geometric theory studying the everyday braid concept, and some generalizations. The idea is that braids can be organised into groups, in which the group operation is 'do the first braid on a set of strings, and then follow it with a second on the twisted strings'. Such groups may be described by explicit presentations, as was shown by Emil Artin (1947). For an elementary treatment along these lines, see the article on braid groups. Braid groups may also be given a deeper mathematical interpretation: as the fundamental group of certain configuration spaces.

Braids as Fundamental Groups

To explain how to reduce a braid group in the sense of Artin to a fundamental group, we consider a connected manifold X of dimension at least 2. The *symmetric product* of n copies of X means the quotient of X^n, the n-fold Cartesian product of X with itself, by the permutation action of the symmetric group on n letters operating on the indices of coordinates. That is, an ordered n-tuple is in the same orbit as any other that is a re-ordered version of it. A path in the n-fold symmetric product is the abstract way of discussing n points of X, considered as an unordered n-tuple, independently tracing out n strings.

Since we must require that the strings never pass through each other, it is necessary that we pass to the subspace Y of the symmetric product, of orbits of n-tuples of *distinct* points. That is, we remove all the subspaces of X^n defined by conditions $x_i = x_j$. This is invariant under the symmetric group, and Y is the quotient by the symmetric group of the non-excluded n-tuples.

Under the dimension condition Y will be connected. With this definition, then, we can call the braid group of X with n strings the fundamental group of Y (for any choice of base point – this is well-defined up to isomorphism).

The case where X is the Euclidean plane is the original one of Artin. In some cases it can be shown that the higher homotopy groups of Y are trivial.

Closed Braids

When X is the plane, the braid can be *closed*, i.e., corresponding ends can be connected in pairs, to form a link, i.e., a possibly intertwined union of possibly knotted loops in three dimensions.

The number of components of the link can be anything from 1 to n, depending on the permutation of strands determined by the link. A theorem of J. W. Alexander demonstrates that every link can be obtained in this way as the "closure" of a braid. Compare with string links.

Different braids can give rise to the same link, just as different crossing diagrams can give rise to the same knot. Markov (1935) describes two moves on braid diagrams that yield equivalence in the corresponding closed braids.

A single-move version of Markov's theorem, was published by Lambropoulou & Rourke (1997). Vaughan Jones originally defined his polynomial as a braid invariant and then showed that it depended only on the class of the closed braid.

Applications

Braid theory has recently been applied to fluid mechanics, specifically to the field of chaotic mixing in fluid flows. The braiding of (2+1) dimensional space-time trajectories formed by motion of physical rods, periodic orbits or "ghost rods", and almost-invariant sets has been used to estimate the topological entropy of several engineered and naturally occurring fluid systems, via the use of Nielsen–Thurston classification.

Braid Group

In mathematics, the braid group on n strands, denoted by B_n, is a group which has an intuitive geometrical representation, and in a sense generalizes the symmetric group S_n.

Here, n is a natural number; if $n > 1$, then B_n is an infinite group. Braid groups find applications in knot theory, since any knot may be represented as the closure of certain braids.

Formal Treatment

To put the above informal discussion of braid groups on firm ground, one needs to use the homotopy concept of algebraic topology, defining braid groups as fundamental groups of a configuration space. This is outlined in the article on braid theory. Alternatively, one can define the braid group purely algebraically via the braid relations, keeping the pictures in mind only to guide the intuition.

History

Braid groups were introduced explicitly by Emil Artin in 1925, although (as Wilhelm Magnus pointed out in 1974) they were already implicit in Adolf Hurwitz's work on monodromy (1891). In fact, as Magnus says, Hurwitz gave the interpretation of a braid group as the fundamental group of a configuration space (cf. braid theory), an interpretation that was lost from view until it was rediscovered by Ralph Fox and Lee Neuwirth in 1962.

Basic Properties

Further Properties

- The braid group B_1 is trivial, B_2 is an infinite cyclic group Z, and B_3 is isomorphic to the knot group of the trefoil knot – in particular, it is an infinite non-abelian group.
- The n-strand braid group B_n embeds as a subgroup into the (n+1)-strand braid group B_{n+1} by adding an extra strand that does not cross any of the first n strands. The increasing union of the braid groups with all $n \geq 1$ is the infinite braid group B_∞.
- All non-identity elements of B_n have infinite order: B_n is torsion-free.
- Patrick Dehornoy constructed a left-invariant linear order on B_n called the Dehornoy order.
- For $n \geq 3$, B_n contains a subgroup isomorphic to the free group on two generators.
- There is a homomorphism $B_n \to$ Z that maps every σ_i to 1. So for instance, the braid $\sigma_2\sigma_3\sigma_1^{-1}\sigma_2\sigma_3$ is mapped to $1 + 1 - 1 + 1 + 1 = 3$.

Interactions of Braid Groups

Relation with Symmetric Group and the Pure Braid Group

By forgetting how the strands twist and cross, every braid on n strands determines a permutation on n elements. This assignment is onto, compatible with composition, and therefore becomes a surjective group homomorphism $B_n \to S_n$ from the braid group into the symmetric group.

The image of the braid

$$\sigma_i \in B_n$$

is the transposition

$$s_i = (i, i+1) \in S_n.$$

These transpositions generate the symmetric group, satisfy the braid group relations, and have order 2. This transforms the Artin presentation of the braid group into the Coxeter presentation of the symmetric group:

$$S_n = \langle s_1, \ldots, s_{n-1} \mid s_i s_{i+1} s_i =$$

$$s_{i+1} s_i s_{i+1}, s_i s_j = s_j s_i \text{ for} \mid \text{i-j} \mid \geq 2, \text{s}_\text{i}^2 = 1 \rangle.$$

The kernel of the homomorphism $B_n \to S_n$ is the subgroup of B_n called the pure braid group on n strands and denoted P_n. In a pure braid, the beginning and the end of each strand are in the like positions. Pure braid groups fit into a short exact sequence

$$1 \to F_{n-1} \to P_n \to P_{n-1} \to 1.$$

This sequence splits and therefore pure braid groups are realised as iterated semi-direct products of free groups.

Relationship to the Mapping Class Group and Classification of Braids

The braid group B_n can be shown to be isomorphic to the mapping class group of a punctured disk with n punctures. This is most easily visualized by imagining each puncture as being connected by a string to the boundary of the disk; each mapping homeomorphism that permutes two of the punctures can then be seen to be a homotopy of the strings, that is, a braiding of these strings. Via this mapping class group interpretation of braids, each braid may be classified as periodic, reducible or pseudo-Anosov.

Connection to Knot Theory and Computational Aspects

If a braid is given and one connects the first left-hand item to the first right-hand item using a new string, the second left-hand item to the second right-hand item etc. (without creating any braids in the new strings), one obtains a link, and sometimes a knot. Alexander's theorem in braid theory states that the converse is true as well: every knot and every link arises in this fashion from at least one braid; such a braid can be obtained by cutting the link. Since braids can be concretely given as words in the generators σ_i, this is often the preferred method of entering knots into computer programs.

The word problem for the braid relations is efficiently solvable and there exists a normal form for elements of B_n in terms of the generators $\sigma_1,...,\sigma_{n-1}$. (In essence, computing the normal form of a braid is the algebraic analogue of "pulling the strands" as illustrated in our second set of images above.)

The free GAP computer algebra system can carry out computations in B_n if the elements are given in terms of these generators. There is also a package called CHEVIE for GAP3 with special support for braid groups. The word problem is also efficiently solved via the Lawrence-Krammer representation. Since there are nevertheless several hard computational problems about braid groups, applications in cryptography have been suggested.

Actions of Braid Groups

In analogy with the action of the symmetric group by permutations, in various mathematical settings there exists a natural action the braid group on n-tuples of objects or on the n-folded tensor product that involves some "twists". Consider an arbitrary group G and let X be the set of all n-tuples of elements of G whose product is the identity element of G. Then B_n acts on X in the following fashion:

$$\sigma_i(x_1,\ldots,x_{i-1},x_i,x_{i+1},\ldots,x_n) = (x_1,\ldots,x_{i-1},x_{i+1},x_{i+1}^{-1}x_ix_{i+1},x_{i+2},\ldots,x_n).$$

Thus the elements x_i and x_{i+1} exchange places and, in addition, x_i is twisted by the inner automorphism corresponding to x_{i+1} — this ensures that the product of the components of x

remains the identity element. It may be checked that the braid group relations are satisfied and this formula indeed defines a group action of B_n on X.

As another example, a braided monoidal category is a monoidal category with a braid group action. Such structures play an important role in modern mathematical physics and lead to quantum knot invariants.

Representations

Elements of the braid group B_n can be represented more concretely by matrices. One classical such representation is Burau representation, where the matrix entries are single variable Laurent polynomials. It had been a long-standing question whether Burau representation was faithful, but the answer turned out to be negative for $n \geq 5$. More generally, it was a major open problem whether braid groups were linear.

In 1990, Ruth Lawrence described a family of more general "Lawrence representations" depending on several parameters. Around 2001 Stephen Bigelow and Daan Krammer independently proved that all braid groups are linear. Their work used the Lawrence–Krammer representation of dimension $n(n-1)/2$ depending on the variables q and t. By suitably specialising these variables, the braid group B_n may be realised as a subgroup of the general linear group over the complex numbers.

Kirby Calculus

In mathematics, the Kirby calculus in geometric topology, named after Robion Kirby, is a method for modifying framed links in the 3-sphere using a finite set of moves, the Kirby moves. Using four-dimensional Cerf theory, he proved that if M and N are 3-manifolds, resulting from Dehn surgery on framed links L and J respectively, then they are homeomorphic if and only if L and J are related by a sequence of Kirby moves.

According to the Lickorish–Wallace theorem any closed orientable 3-manifold is obtained by such surgery on some link in the 3-sphere. Some ambiguity exists in the literature on the precise use of the term "Kirby moves". Different presentations of "Kirby calculus" have a different set of moves and these are sometimes called Kirby moves.

Kirby's original formulation involved two kinds of move, the "blow-up" and the "handle slide"; Roger Fenn and Colin Rourke exhibited an equivalent construction in terms of a single move, the Fenn–Rourke move, that appears in many expositions and extensions of the Kirby calculus. Dale Rolfsen's book, *Knots and Links*, from which many topologists have learned the Kirby calculus, describes a set of two moves: 1) delete or add a component with surgery coefficient infinity 2) twist along an unknotted component and modify surgery coefficients appropriately (this is called the Rolfsen twist). This allows an extension of the Kirby calculus to rational surgeries.

There are also various tricks to modify surgery diagrams. One such useful move is the slam-dunk. An extended set of diagrams and moves are used for describing 4-manifolds. A framed link in the 3-sphere encodes instructions for attaching 2-handles to the 4-ball. (The 3-dimensional boundary of this manifold is the 3-manifold interpretation of the link diagram mentioned above.) 1-handles are denoted by either (a) a pair of 3-balls (the attaching region of the 1-handle) or, more commonly, (b) unknotted circles with dots. The dot indicates that a neighbourhood of a standard 2-disk with boundary the dotted circle is to be excised from the interior of the 4-ball. Excising this 2-handle is equivalent to adding a 1-handle. 3-handles and 4-handles are usually not indicated in the diagram.

Chapter 8

Real Projective Plane

In mathematics, the real projective plane is an example of a compact non-orientable two-dimensional manifold, that is, a one-sided surface. It cannot be embedded in our usual three-dimensional space without intersecting itself. It has basic applications to geometry, since the common construction of the real projective plane is as the space of lines in R^3 passing through the origin.

The plane is also often described topologically, in terms of a construction based on the Möbius strip: if one could glue the (single) edge of the Möbius strip to itself in the correct direction, one would obtain the projective plane. (This cannot be done in our three-dimensional space.) Equivalently, gluing a disk along the boundary of the Möbius strip gives the projective plane. Topologically, it has Euler characteristic 1, hence a demigenus (non-orientable genus, Euler genus) of 1. Since the Möbius strip, in turn, can be constructed from a square by gluing two of its sides together, the real projective plane can thus be represented as a unit square (that is, $[0,1] \times [0,1]$) with its sides identified by the following equivalence relations:

$$(0, y) \sim (1, 1-y) \quad \text{for } 0 < y \leq 1$$

and

$$(x, 0) \sim (1-x, 1) \quad \text{for } 0 \leq x \leq 1,$$

as in the diagram on the right.

Examples : Topology is not concerned with flatness, and the real projective plane may be twisted up and placed in the Euclidean plane or 3-space in many different ways. Some of the more important examples are described below. The projective plane cannot be embedded (that is without intersection) in three-dimensional Euclidean space.

The proof that the projective plane does not embed in three-dimensional Euclidean space goes like this: Assume that it does embed, it would bound a compact region in three-dimensional Euclidean space by the generalized Jordan curve theorem. The outward-pointing unit normal vector field would then give an orientation of the boundary manifold, but the boundary manifold would be projective space, which is not orientable. This is a contradiction, and so our assumption that it does embed must have been false.

The Projective Sphere

Consider a sphere, and let the great circles of the sphere be "lines", and let pairs of antipodal points be "points". It is easy to check that this system obeys the axioms required of a projective plane:

- any pair of distinct great circles meet at a pair of antipodal points; and
- any two distinct pairs of antipodal points lie on a single great circle.

If we identify each point on the sphere with its antipodal point, then we get a representation of the real projective plane in which the "points" of the projective plane really are points. This means that the projective plane is the quotient space of the sphere obtained by partitioning the sphere into equivalence classes under the equivalence relation ~, where x ~ y if y = "x. This quotient space of the sphere is homeomorphic with the collection of all lines passing through the origin in R^3.

The quotient map from the sphere onto the real projective plane is in fact a two sheeted (i.e. two-to-one) covering map. It follows that the fundamental group of the real projective plane is the cyclic group of order 2, i.e. integers modulo 2. One can take the loop *AB* from the figure above to be the generator.

The Projective Hemisphere

Because the real projective plane covers the sphere twice, it may be represented as a hemisphere around whose rim opposite points are similarly identified.

Boy's Surface – an Immersion

The projective plane can be immersed (local neighbourhoods do not have self-intersections) in 3-space. Boy's surface is an example of an immersion. Polyhedral examples must have at least nine faces.

Roman Surface

Steiner's Roman surface is a more degenerate map of the projective plane into 3-space, containing a cross-cap.

A polyhedral representation is the tetrahemihexahedron, which has the same general form as Steiner's Roman Surface, shown to the right.

Hemi Polyhedra

Looking in the opposite direction, certain abstract regular polytopes — hemi-cube, hemi-dodecahedron, and hemi-icosahedron — can be constructed as regular figures in the *projective plane.*

Planar Projections

Various planar (flat) projections or mappings of the projective plane have been described. In 1874 Klein described the mapping $k(x,y)=(1+x^2+y^2)^{1/2}.(x,y)$

Central projection of the projective hemisphere onto a plane yields the usual infinite projective plane, described below.

Homogeneous Coordinates

The points in the plane can be represented by homogeneous coordinates. A point has homogeneous coordinates $[x:y:z]$, where the coordinates $[x:y:z]$ and $[tx:ty:tz]$ are considered to represent the same point, for all nonzero values of t.

The points with coordinates $[x:y:1]$ are the usual real plane, called the finite part of the projective plane, and points with coordinates $[x:y:0]$, called points at infinity or ideal points, constitute a line called the line at infinity. (The homogeneous coordinates $[0:0:0]$ do not represent any point.)

The lines in the plane can also be represented by homogeneous coordinates. A projective line corresponding to the plane $ax+by+c=0$ in R^3 has the homogeneous coordinates $(a:b:c)$. Thus, these coordinates have the equivalence relation

$(a : b : c) = (da : db : dc)$ for all nonzero values of d. Hence a different equation of the same line $dax + dby + dc = 0$ gives the same homogeneous coordinates. A point $[x : y : z]$ lies on a line $(a : b : c)$ if $ax + by + c = 0$. Therefore, lines with coordinates $(a : b : c)$ where a, b are not both 0 correspond to the lines in the usual real plane, because they contain points that are not at infinity. The line with coordinates $(0 : 0 : 1)$ is the line at infinity, since the only points on it are those with $z = 0$.

The Flat Projective Plane

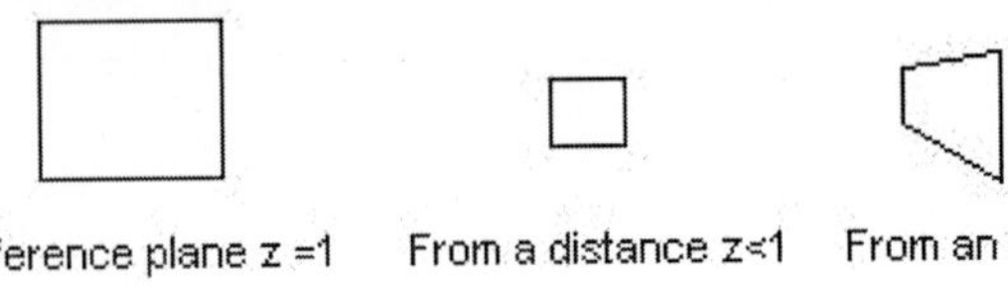

Table: In the projective plane P^2, a point x is represented by the homogeneous coordinate (x_1, x_2, x_3). If we think of (x_1, x_2, x_3) as a point in real space R^3 with the third value of the homogeneous coordinate as a value in the z direction, then P^2 can be visualized as R^3.

Points, Rays, Lines, and Planes

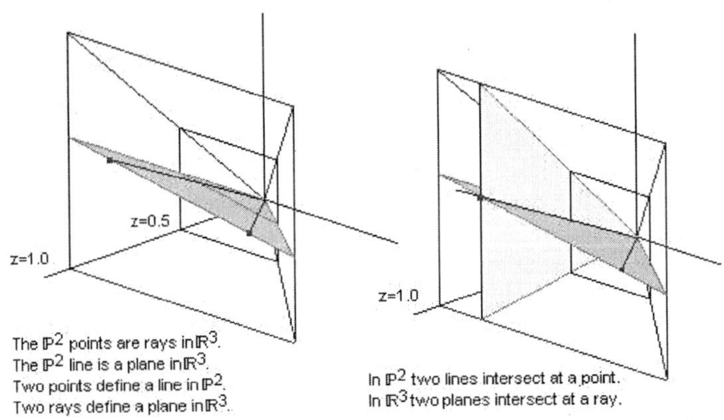

Table: A line in P^2 can be represented by the equation $ax + by + c = 0$. If we treat a, b, and c as the column vector –! and x, y, 1 as the column vector x then the equation above can be written in matrix form as:

$$\mathbf{x}^T \ell = 0 \text{ or } \ell^T \mathbf{x} = 0.$$

Using vector notation we may instead write

$$\mathbf{x} \cdot \ell = 0 \text{ or } \ell \cdot \mathbf{x} = 0.$$

The equation $k(x^T–!) = 0$ (which k is a non-zero scalar) sweeps out a plane that goes through zero in R^3 and $k(x)$ sweeps out a ray, again going through zero. The plane and ray are subspaces in R^3, which always go through zero.

Ideal Points

If $z = 1$ we have a normalized homogeneous coordinate. All points that have $z = 1$ create a plane. Let's pretend we are looking at that plane (from a position further out along the z axis and looking back towards the origin) and there are two parallel lines drawn on the plane. From where we are standing (given our visual capabilities) we can see only so much of the plane, which we represent as the area outlined in red in the diagram.

If we walk away from the plane along the z axis, (still looking backwards towards the origin), we can see more of the plane. In our field of view original points have moved. We can reflect this movement by dividing the homogeneous coordinate by a constant. In the image to the right we have divided by 2 so the z value now becomes 0.5. If we walk far enough away what we are looking at becomes a point in the distance. As we walk away we see more and more of the parallel lines. The lines will meet at a line at infinity (a line that goes through zero on the plane at $z = 0$). Lines on the plane when $z = 0$ are ideal points. The plane at $z = 0$ is the line at infinity.

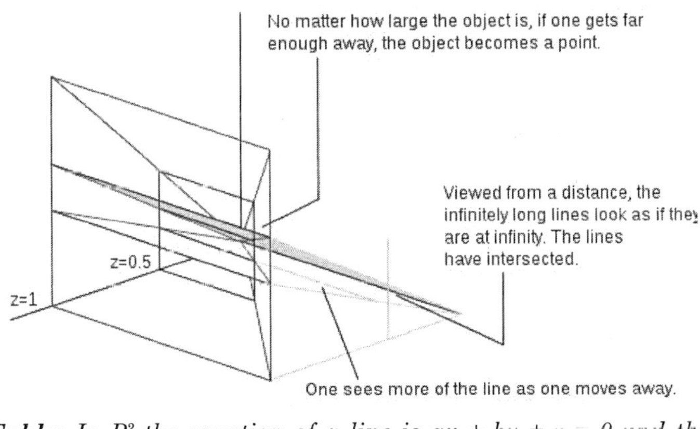

Table: *In P^2 the equation of a line is $ax + by + c – 0$ and this equation can represent a line on any plane parallel to the x, y plane by multiplying the equation by k.*

The homogeneous point (0, 0, 0) is where all the real points go when you're looking at the plane from an infinite distance, a line on the $z = 0$ plane is where parallel lines intersect.

Duality

In the equation $\mathbf{x}^T \ell = 0$ there are two column vectors. You can keep either constant and vary the other. If we keep the point constant x and vary the coefficients ℓ we create new lines that go through the point.

If we keep the coefficients constant and vary the points that satisfy the equation we create a line. We look upon x as a point because the axes we are using are x, y, and z. If we instead plotted the coefficients using axis marked a, b, c points would become lines and lines would become points. If you prove something with the data plotted on axis marked x, y, and z the same argument can be used for the data plotted on axis marked a, b, and c. That is duality.

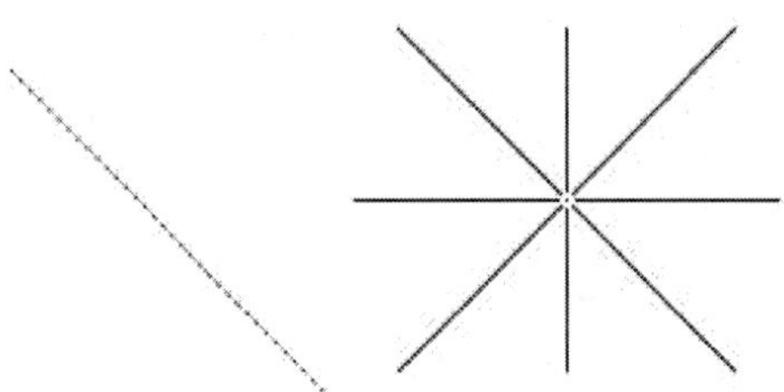

As you vary , youx create points that trace out a line.

As you vary , youtl create lines that go through the fixed point.

Lines Joining Points and Intersection of Lines (Using Duality)

The equation $x^T \ell = 0$ calculates the inner product of two column vectors. The inner product of two vectors is zero if the vectors are orthogonal. To find the line between the points x_1 and x_2 you must find the column vector ℓ that satisfies the equations $x_1{}^T \ell = 0$ and $x_2{}^T \ell = 0$, that is we must find a column vector ℓ that is orthogonal to x_1 and x_2.

In the case of P^2, the cross product will find such a vector. The line joining two points is given by the equation $x_1 \times x_2$. To find the intersection of two lines you look to duality. If you plot –! in the coefficient space you get rays. To find the point x that is orthogonal to the two rays you find the cross product. That is $\ell_1 \times \ell_2$. While the cross product works in P^2, it is not well-defined in arbitrary dimensions. However, this pair of equations is satisfied by

$$\mathbf{x}_1^{\mathrm{T}}\ell - \lambda \mathbf{x}_2^{\mathrm{T}}\ell = 0$$

Higher Non-Orientable Surfaces

By glueing together projective planes successively we get non-orientable surfaces of higher demigenus. The glueing process consists of cutting out a little disk from each surface and identifying (*glueing*) their boundary circles. Glueing two projective planes creates the Klein bottle. The article on the fundamental polygon describes the higher non-orientable surfaces.

Annulus (Mathematics)

The open annulus is topologically equivalent to both the open cylinder $S^1 \times (0,1)$ and the punctured plane.

The area of an annulus is the difference in the areas of the larger circle of radius R and the smaller one of radius r:

$$A = \pi R^2 - \pi r^2 = \pi(R^2 - r^2).$$

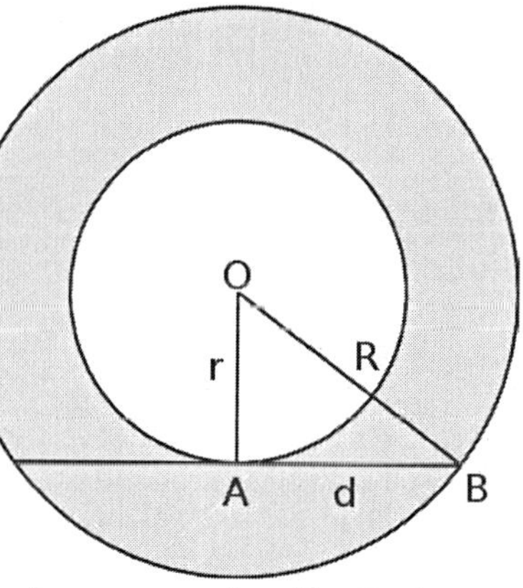

Table: *In mathematics, an annulus (the Latin word for "little ring", with plural* annuli*) is a ring-shaped geometric figure, or more generally, a term used to name a ring-shaped object. Or, it is the area between two concentric circles. The adjectival form is annular (for example, an annular eclipse).*

The area of an annulus can be obtained from the length of the longest interval that can lie completely inside the annulus, 2*d in the accompanying diagram. This can be proven by the Pythagorean theorem; the length of the longest interval that can lie completely inside the annulus will be tangent to the smaller circle and form a right angle with its radius at that point. Therefore d and r are the sides of a right angled triangle with hypotenuse R and the area is given by:

$$A = \pi(R^2 - r^2) = \pi d^2.$$

The area can also be obtained via calculus by dividing the annulus up into an infinite number of annuli of infinitesimal width $d\rho$ and area $2\pi\rho\, d\rho$ ( ) and then integrating from $\rho = r$ to $\rho = R$:

$$A = \int_r^R 2\pi\rho d\rho = \pi(R^2 - r^2).$$

The area of an annulus segment of angle θ, with θ measured in radians, is given by:

$$A = \frac{\theta}{2}(R^2 - r^2)$$

Complex Structure

In complex analysis an annulus ann$(a; r, R)$ in the complex plane is an open region defined by:

$$r < |z - a| < R.$$

If r is 0, the region is known as the punctured disk of radius R around the point a.

As a subset of the complex plane, an annulus can be considered as a Riemann surface. The complex structure of an annulus depends only on the ratio r/R. Each annulus ann$(a; r, R)$ can be holomorphically mapped to a standard one centred at the origin and with outer radius 1 by the map

$$z \mapsto \frac{z - a}{R}.$$

The inner radius is then $r/R < 1$.

The Hadamard three-circle theorem is a statement about the maximum value a holomorphic function may take inside an annulus.

Torus

In geometry, a torus (pl. tori) is a surface of revolution generated by revolving a circle in three dimensional space about an axis coplanar with the circle. If the axis of revolution does not touch the circle, the surface has a ring shape and is called a ring torus or simply *torus* if the ring shape is implicit.

When the axis is tangent to the circle, the resulting surface is called a horn torus; when the axis is a chord of the circle, it is called a spindle torus. A degenerate case is when the axis is a diametre of the circle, which simply generates the surface of a sphere.

The ring torus bounds a solid known as a toroid. The adjective toroidal can be applied to tori, toroids or, more generally, any ring shape as in toroidal inductors and transformers. Real-world examples of (approximately) toroidal objects include doughnuts, vadais, inner tubes, many lifebuoys, O-rings and vortex rings.

In topology, a ring torus is homeomorphic to the Cartesian product of two circles: $S^1 \times S^1$, and the latter is taken to be the definition in that context. It is a compact 2-manifold of genus 1. The ring torus is one way to embed this space into Euclidean space, but another way to do this is the Cartesian product of the embedding of S^1 in the plane. This produces a geometric object called the Clifford torus, surface in 4-space.

The word *torus* comes from the Latin word meaning cushion.

Geometry

A torus can be defined parametrically by:

$$x(u,v) = (R + r\cos v)\cos u$$

$$y(u,v) = (R + r\cos v)\sin u$$

$$z(u,v) = r\sin v$$

where

u,v are in the interval $[0, 2\pi)$,

R (or A) is the distance from the centre of the tube to the centre of the torus,

r (or a) is the radius of the tube.

R and r are also known as the "major radius" and "minor radius", respectively. The ratio of the two is known as the "aspect ratio". A doughnut has an aspect ratio of 2 to 3.

An implicit equation in Cartesian coordinates for a torus radially symmetric about the z-axis is

$$\left(R-\sqrt{x^2+y^2}\right)^2+z^2=r^2,$$

or the solution of $f(x,y,z)=0$, where

$$f(x,y,z)=\left(R-\sqrt{x^2+y^2}\right)^2+z^2-r^2.$$

Algebraically eliminating the square root gives a quartic equation,

$$(x^2+y^2+z^2+R^2-r^2)^2=4R^2(x^2+y^2).$$

The three different classes of standard tori correspond to the three possible relative sizes of r and R. When $R > r$, the surface will be the familiar ring torus. The case $R = r$ corresponds to the horn torus, which in effect is a torus with no "hole".The case $R < r$ describes the self-intersecting spindle torus. When $R = 0$, the torus degenerates to the sphere. The surface area and interior volume of this torus are easily computed using Pappus's centroid theorem giving.

$$A=4\pi^2Rr=(2\pi r)(2\pi R)$$
$$V=2\pi^2Rr^2=(\pi r^2)(2\pi R).$$

These formulas are the same as for a cylinder of length $2\pi R$ and radius r, created by cutting the tube and unrolling it by straightening out the line running around the centre of the tube.

The losses in surface area and volume on the inner side of the tube exactly cancel out the gains on the outer side. As a torus is the product of two circles, a modified version of the spherical coordinate system is sometimes used.

In traditional spherical coordinates there are three measures, R, the distance from the centre of the coordinate system, and θ ϕ, angles measured from the centre point. As a torus has, effectively, two centre points, the centrepoints of the angles are moved; measures the same angle as it does in the spherical system, but is known as the "toroidal" direction. The centre point of is moved to the centre of r, and is known as the "poloidal" direction.

These terms were first used in a discussion of the Earth's magnetic field, where "poloidal" was used to denote "the direction toward the poles". In modern use these terms are more commonly used to discuss magnetic confinement fusion devices.

Topology

Topologically, a torus is a closed surface defined as the product of two circles: $S^1 \times S^1$. This can be viewed as lying in C^2 and is a subset of the 3-sphere S^3 of radius "2. This topological torus is also often called the Clifford torus. In fact, S^3 is filled out by a family of nested tori in this manner (with two degenerate circles), a fact which is important in the study of S^3 as a fibre bundle over S^2 (the Hopf bundle).

The surface described above, given the relative topology from R^3, is homeomorphic to a topological torus as long as it does not intersect its own axis. A particular homeomorphism is given by stereographically projecting the topological torus into R^3 from the north pole of S^3.

The torus can also be described as a quotient of the Cartesian plane under the identifications

$$(x,y) \sim (x+1,y) \sim (x,y+1).$$

Or, equivalently, as the quotient of the unit square by pasting the opposite edges together, described as a fundamental polygon $ABA^{-1}B^{-1}$. The fundamental group of the torus is just the direct product of the fundamental group of the circle with itself:

$$\pi_1(\mathbb{T}^2) = \pi_1(S^1) \times \pi_1(S^1) \cong \mathbb{Z} \times \mathbb{Z}.$$

Intuitively speaking, this means that a closed path that circles the torus' "hole" (say, a circle that traces out a particular latitude) and then circles the torus' "body" (say, a circle that traces out a particular longitude) can be deformed to a path that circles the body and then the hole.

So, strictly 'latitudinal' and strictly 'longitudinal' paths commute. This might be imagined as two shoelaces passing through each other, then unwinding, then rewinding.

If a torus is punctured and turned inside out then another torus results, with lines of latitude and longitude interchanged. The first homology group of the torus is isomorphic to the fundamental group (this follows from Hurewicz theorem since the fundamental group is abelian).

Two-Sheeted Cover

The 2-torus double-covers the 2-sphere, with four ramification points. Every conformal structure on the 2-torus can be represented as a two-sheeted cover of the 2-sphere. The points on the torus corresponding to the ramification points are the Weierstrass points. In fact, the conformal type of the torus is determined by the cross-ratio of the four points.

n-Dimensional Torus

The torus has a generalization to higher dimensions, the n-dimensional torus, often called the n-torus for short. (This is one of two different meanings of the term "n-torus".) Recalling that the torus is the product space of two circles, the n-dimensional torus is the product of n circles. That is:

$$\mathbb{T}^n = \underbrace{S^1 \times S^1 \times \cdots \times S^1}_{n}.$$

The torus discussed above is the 2-dimensional torus. The 1-dimensional torus is just the circle. Just as for the 2-torus, the n-torus can be described as a quotient of R^n under integral shifts in any coordinate. That is, the n-torus is R^n modulo the action of the integer lattice Z^n (with the action being taken as vector addition). Equivalently, the n-torus is obtained from the n-dimensional hypercube by gluing the opposite faces together.

An n-torus in this sense is an example of an n-dimensional compact manifold. It is also an example of a compact abelian Lie group. This follows from the fact that the unit circle is a compact abelian Lie group (when identified with the unit complex numbers with multiplication). Group multiplication on the torus is then defined by coordinate-wise multiplication. Toroidal groups play an important part in the theory of compact Lie groups. This is due in part to the fact that in any compact Lie group G one can always find a maximal torus; that is, a closed subgroup which is a torus of the largest possible dimension. Such maximal tori T have a controlling role to play in theory of connected G.

Automorphisms of T are easily constructed from automorphisms of the lattice Z^n, which are classified by integral matrices M of size $n \times n$ which are invertible with integral inverse; these are just the integral M of determinant +1 or −1. Making M act on R^n in the usual way, one has the typical toral automorphism on the quotient.

The fundamental group of an n-torus is a free abelian group of rank n. The k-th homology group of an n-torus is a free abelian group of rank n choose k. It follows that the Euler characteristic of the n-torus is 0 for all n. The cohomology ring $H^{\cdot}(T^n,Z)$ can be identified with the exterior algebra over the Z-module Z^n whose generators are the duals of the n nontrivial cycles.

Flat Torus

The flat torus is a specific embedding of the familiar 2-torus into Euclidean 4-space or higher dimensions. Its surface has zero Gaussian curvature everywhere. Its surface is "flat" in the same sense that the surface of a cylinder is "flat". In 3 dimensions one can bend a flat sheet of paper into a cylinder without stretching the paper, but you cannot then bend this cylinder into a torus without stretching the paper (unless you give up some regularity and differentiability conditions). In 4 dimensions one can (mathematically).

A simple 4-d Euclidean embedding is as follows: $<x,y,z,w> = <R \cos u, R \sin u, P \cos v, P \sin v>$ where R and P are constants determining the aspect ratio. It is diffeomorphic to a regular torus but not isometric. It can not be isometrically embedded into Euclidean 3-space. Mapping it into 3-space requires you to stretch it, in which case it looks like a regular torus, for example, the following map $<x,y,z> = <(R + P \sin v)\cos u, (R + P \sin v)\sin u, P \cos v>$.

A flat torus partitions the 3-sphere into two congruent solid tori subsets with the aforesaid flat torus surface as their common boundary. Recently (April 2012), an embedding of a flat torus into three dimensions was found. It is similar in structure to a fractal as it is constructed by repeatedly corrugating a normal torus. Like fractals, it has no defined

Gaussian curvature. However, unlike fractals, it does have defined surface normals.

The theory of surfaces the term n-torus has a different meaning. Instead of the product of n circles, they use the phrase to mean the connected sum of n 2-dimensional tori.

To form a connected sum of two surfaces, remove from each the interior of a disk and "glue" the surfaces together along the disks' boundary circles.

To form the connected sum of more than two surfaces, sum two of them at a time until they are all connected. In this sense, an n-torus resembles the surface of n doughnuts stuck together side by side, or a 2-dimensional sphere with n handles attached.

An ordinary torus is a 1-torus, a 2-torus is called a double torus, a 3-torus a triple torus, and so on. The n-torus is said to be an "orientable surface" of "genus" n, the genus being the number of handles. The 0-torus is the 2-dimensional sphere.

The classification theorem for surfaces states that every compact connected surface is either a sphere, an n-torus with $n > 0$, or the connected sum of n projective planes (that is, projective planes over the real numbers) with $n > 0$.

Toroidal Polyhedra

Polyhedra with the topological type of a torus are called toroidal polyhedra, and satisfy a modified version of the polyhedron formula, $E - F - V = 0$.

The term "toroidal polydron" is also used for higher genus polyhedra and for immersions of toroidal polyhedra.

Automorphisms

The homeomorphism group (or the subgroup of diffeomorphisms) of the torus is studied in geometric topology.

Its mapping class group (the group of connected components) is isomorphic to the group GL(n, Z) of invertible integer matrices, and can be realised as linear maps on the universal covering space $\mathbf{R}^n$ that preserve the standard lattice $\mathbf{Z}^n$ (this corresponds to integer coefficients) and thus descend to the quotient.

At the level of homotopy and homology, the mapping class group can be identified as the action on the first homology (or equivalently, first cohomology, or on the fundamental group, as these are all naturally isomorphic; note also that the first cohomology group generates the cohomology algebra):

$$\mathrm{MCG}(T^n) = \mathrm{Aut}(\pi_1(X)) = \mathrm{Aut}(\mathbf{Z}^n) = \mathrm{GL}(n, \mathbf{Z}).$$

Since the torus is an Eilenberg-MacLane space $K(G, 1)$, its homotopy equivalences, up to homotopy, can be identified with automorphisms of the fundamental group); that this agrees with the mapping class group reflects that all homotopy equivalences can be realised by homeomorphisms – every homotopy equivalence is homotopic to a homeomorphism – and that homotopic homeomorphisms are in fact isotopic (connected through homeomorphisms, not just through homotopy equivalences).

More tersely, the map $\mathrm{Homeo}(T^n) \to \mathrm{SHE}(T^n)$ is 1-connected (isomorphic on path-components, onto fundamental group). This is a "homeomorphism reduces to homotopy reduces to algebra" result.

Thus the short exact sequence of the mapping class group splits (an identification of the torus as the quotient of $\mathbf{R}^n$ gives a splitting, via the linear maps, as above):

$$1 \to \mathrm{Homeo}_0(T^n) \to \mathrm{Homeo}(T^n) \to \mathrm{MCG}(T^n) \to 1,$$

so the homeomorphism group of the torus is a semidirect product,

$$\mathrm{Homeo}(T^n) \cong \mathrm{Homeo}_0(T^n) \rtimes \mathrm{GL}(n, \mathbf{Z}).$$

The mapping class group of higher genus surfaces is much more complicated, and an area of active research.

Colouring a Torus

If a torus is divided into regions, then it is always possible to colour the regions with no more than seven colours so that neighbouring regions have different colours. (Contrast with the four colour theorem for the plane.)

Cutting a Torus

A standard torus (specifically, a ring torus) can be cut with n planes into at most

$$\frac{1}{6}(n^3 + 3n^2 + 8n) \text{parts.}$$

The initial terms of this sequence for n starting from 1 are:

2, 6, 13, 24, 40, ... (sequence A003600 in OEIS).

Klein Bottle

In mathematics, the Klein bottle is a non-orientable surface, informally, a surface (a two-dimensional manifold) in which notions of left and right cannot be consistently defined. Other related non-orientable objects include the Möbius strip and the real projective plane.

Whereas a Möbius strip is a surface with boundary, a Klein bottle has no boundary. (For comparison, a sphere is an orientable surface with no boundary.)

The Klein bottle was first described in 1882 by the German mathematician Felix Klein. It may have been originally named the *Kleinsche Fläche* ("Klein surface") and that this was incorrectly interpreted as *Kleinsche Flasche* ("Klein bottle"), which ultimately led to the adoption of this term in the German language as well.

Properties

The Klein bottle can be seen as a fibre bundle over S^1 as follows: one takes the square (modulo the edge identifying equivalence relation) from above to be E, the total space, while the base space B is given by the unit interval in y, modulo $1\sim 0$. The projection π is then given by $\pi([x, y]) = [y]$. Like the Möbius strip, the Klein bottle is a two-dimensional differentiable manifold which is not orientable.

Unlike the Möbius strip, the Klein bottle is a *closed* manifold, meaning it is a compact manifold without boundary. While the Möbius strip can be embedded in three-dimensional Euclidean space R^3, the Klein bottle cannot. It can be embedded in R^4, however.

The Klein bottle can be constructed (in a mathematical sense, because it cannot be done without allowing the surface to intersect itself) by joining the edges of two Möbius strips together, as described in the following anonymous limerick:

A mathematician named Klein
Thought the Möbius band was divine.
Said he: "If you glue
The edges of two,
You'll get a weird bottle like mine."

Six colours suffice to colour any map on the surface of a Klein bottle; this is the only exception to the Heawood conjecture, a generalization of the four colour theorem, which would require seven. A Klein bottle is equivalent to a sphere plus two cross caps.

When embedded in Euclidean space the Klein bottle is one-sided. However there are other topological 3-spaces, and in some of the non-orientable examples a Klein bottle can be embedded such that it is two-sided, though due to the nature of the space it remains non-orientable.

Dissection

Dissecting a Klein bottle into halves along its plane of symmetry results in two mirror image Möbius strips, i.e. one with a left-handed half-twist and the other with a right-handed half-twist (one of these is pictured on the right). Remember that the intersection pictured isn't really there.

Simple-Closed Curves

One description of the types of simple-closed curves that may appear on the surface of the Klein bottle is given by the use of the first homology group of the Klein bottle calculated with integer coefficients.

This group is isomorphic to $Z \times Z_2$. Up to reversal of orientation, the only homology classes which contain simple-closed curves are as follows: (0,0), (1,0), (1,1), (2,0), (0,1).

Up to reversal of the orientation of a simple closed curve, if it lies within one of the two crosscaps that make up the Klein bottle, then it is in homology class (1,0) or (1,1); if it cuts the Klein bottle into two Möbius bands, then it is in homology class (2,0); if it cuts the Klein bottle into an annulus, then it is in homology class (0,1); and if bounds a disk, then it is in homology class (0,0).

Parameterization

The "figure" immersion (Klein bagel) of the Klein bottle has a particularly simple parameterization. It is that of a "figure– torus with a 180 degree "Möbius" twist inserted:

$$x = \left(r + \cos\frac{u}{2}\sin v - \sin\frac{u}{2}\sin 2v\right)\cos u$$

$$y = \left(r + \cos\frac{u}{2}\sin v - \sin\frac{u}{2}\sin 2v\right)\sin u$$

$$z = \sin\frac{u}{2}\sin v + \cos\frac{u}{2}\sin 2v$$

In this immersion, the self-intersection circle is a geometric circle in the xy-plane. The positive constant r is the radius of this circle.

The parameter u gives the angle in the xy-plane, and v specifies the position around the 8-shaped cross section. With the above parameterization the cross section is a 2:1 Lissajous curve.

The parameterization of the 3-dimensional immersion of the bottle itself is much more complicated. Here is a simplified version:

$$x = \frac{\sqrt{2}f(u)\cos u\cos v(3\cos^2 u - 1) - 2\cos 2u}{80\pi^3 g(u)} - \frac{3\cos u - 3}{4}$$

$$y = -\frac{f(u)\sin v}{60\pi^3}$$

$$z = -\frac{\sqrt{2}f(u)\sin u\cos v}{15\pi^3 g(u)} + \frac{\sin u\cos^2 u + \sin u}{4} - \frac{\sin u\cos u}{2}$$

where

$$f(u) = 20u^3 - 65\pi u^2 + 50\pi^2 u - 16\pi^3$$

$$g(u) = \sqrt{\begin{matrix}8\cos^2 2u - \cos 2u(24\cos^3 u - 8\cos u + 15) \\ +6\cos^4 u(1 - 3\sin^2 u) + 17\end{matrix}}$$

for $0 \le u < 2\pi$ and $0 \le v < 2\pi$.

In this parameterization, u follows the length of the bottle's body while v goes around its circumference.

Generalizations

The generalization of the Klein bottle to higher genus is given in the article on the fundamental polygon. In another order of ideas, constructing 3-manifolds, it is known that a solid Klein bottle is topologically equivalent with the Cartesian product: $Mo \times I$, the Mobius band times an interval. The solid Klein bottle is the non-orientable version of the solid torus, equivalent to $D^2 \times S^1$.

Klein Surface

A Klein surface is, as for Riemann surfaces, a surface with an atlas allowing the transition functions to be composed using complex conjugation. One can obtain the so called dianalytic structure of the space.

Boy's Surface

In geometry, Boy's surface is an immersion of the real projective plane in 3-dimensional space found by Werner Boy in 1901 (he discovered it on assignment from David Hilbert to prove that the projective plane *could not* be immersed in 3-space). Unlike the Roman surface and the cross-cap, it has no singularities (i.e., pinch-points), but it does self-intersect.

To make a Boy's surface:

1. Start with a sphere. Remove a cap.
2. Attach one end of each of three strips to alternate sixths of the edge left by removing the cap.
3. Bend each strip and attach the other end of each strip to the sixth opposite the first end, so that the inside of the sphere at one end is connected to the outside at the other. Make the strips skirt the middle rather than go through it.
4. Join the loose edges of the strips. The joins intersect the strips.

Boy's surface is discussed (and illustrated) in Jean-Pierre Petit's *Le Topologicon*. Boy's surface was first parametrized explicitly by Bernard Morin in 1978.

Boy's surface is one of the two possible immersions of the real projective plane which have only a single triple point.

Symmetry of the Boy's Surface

Boy's surface has 3-fold symmetry. This means that it has an axis of discrete rotational symmetry: any 120° turn about this axis will leave the surface looking exactly the same. The Boy's surface can be cut into three mutually congruent pieces.

Model at Oberwolfach

The Mathematical Research Institute of Oberwolfach has a large model of a Boy's surface outside the entrance, constructed and donated by Mercedes-Benz in January 1991. This model has 3-fold rotational symmetry and minimises the Willmore energy of the surface.

It consists of steel strips which represent the image of a polar coordinate grid under a parameterization given by Robert Bryant and Rob Kusner. The meridians (rays) become ordinary Möbius strips, i.e. twisted by 180 degrees.

All but one of the strips corresponding to circles of latitude (radial circles around the origin) are untwisted, while the one corresponding to the boundary of the unit circle is a Möbius strip twisted by three times 180 degrees — as is the emblem of the institute (Mathematisches Forschungsinstitut Oberwolfach 2008).

Applications

Boy's surface can be used in sphere eversion, as a half-way model. A half-way model is an immersion of the sphere with the property that a rotation interchanges inside and outside, and so can be employed to evert (turn inside-out) a sphere.

Boy's (the case p = 3) and Morin's (the case p = 2) surfaces are beginning of a sequence of half-way models with higher symmetry first proposed by George Francis, indexed by the even integers 2p (for p odd, these immersions can be factor through a projective plane). Kusner's parametrization yields all these.

Parametrization of Boy's Surface

Boy's surface can be parametrized in several ways. One parametrization, discovered by Rob Kusner and Robert Bryant, is the following: given a complex number z whose magnitude is less than or equal to one ($\| z \| \leq 1$), let

$$g_1 = -\frac{3}{2}\text{Im}\left[\frac{z(1-z^4)}{z^6+\sqrt{5}z^3-1}\right]$$

$$g_2 = -\frac{3}{2}\text{Re}\left[\frac{z(1+z^4)}{z^6+\sqrt{5}z^3-1}\right]$$

$$g_3 = \text{Im}\left[\frac{1+z^6}{z^6+\sqrt{5}z^3-1}\right]-\frac{1}{2}$$

so that

$$\begin{pmatrix} x \\ y \\ z \end{pmatrix} = \frac{1}{g_1^2+g_2^2+g_3^2}\begin{pmatrix} g_1 \\ g_2 \\ g_3 \end{pmatrix}$$

where x, y, and z are the desired Cartesian coordinates of a point on the Boy's surface. where x, y, and z are the desired Cartesian coordinates of a point on the Boy's surface.

If one performs an inversion of this parametrization centred on the triple point, one obtains a complete minimal surface with three "ends" (that's how this parametrization was discovered naturally). This implies that the Bryant-Kusner parametrization of Boy's surfaces is "optimal" in the sense that it is the "least bent" immersion of a projective plane into three-space.

Property of Bryant-Kusner Parametrization

Further information: Properties of the Bryant-Kusner parametrization

If z is replaced by the negative reciprocal of its complex conjugate, $-\frac{1}{z^*}$, then the functions g_1, g_2, and g_3 of z are left unchanged.

Relating the Boy's Surface to the Real Projective Plane

Let $P(z) = (x(z), y(z), z(z))$, where $\| z \| \leq 1$, denote a point on Boy's surface. Then

$$P(z) = P\left(-\frac{1}{z^*}\right)$$

but only if $\| z \| = \sqrt{zz^{\star}} = 1$. What if $\| z \| < 1$? Then

$$\left\| -\frac{1}{z^{\star}} \right\| > 1$$

because

$$-\frac{1}{z^{\star}} = \frac{-z}{z^{\star}z} = \frac{-z}{\| z \|^2}$$

whose magnitude is

$$\frac{\| z \|}{\| z \|^2} = \frac{1}{\| z \|},$$

but $\| z \| < 1$, so that

$$\frac{1}{\| z \|} > 1.$$

Since $P(z)$ belongs to the Boy's surface only when $\| z \| \leq 1$,, this means that $P\left(-\frac{1}{z^{\star}}\right)$ belongs to Boy's surface only if $\| z \| = 1$. Thus $P(z) = P(-z)$ if $\| z \| = 1$, but all other points on the Boy's surface are unique. The set of complex values $\| z \| \leq 1$ is the unit disk. Thus the Boy's surface has been parametrized by a disk such that pairs of diametrically opposite points on the perimeter of the disk are equivalent. Therefore the Boy's surface is homeomorphic to the real projective plane, RP^2.

Roman Surface

The Roman surface or Steiner surface (so called because Jakob Steiner was in Rome when he thought of it) is a self-intersecting mapping of the real projective plane into three-dimensional space, with an unusually high degree of symmetry. This mapping is not an immersion of the projective plane; however the figure resulting from removing six singular points is one.

The simplest construction is as the image of a sphere centred at the origin under the map $f(x,y,z) = (yz,xz,xy)$. This gives an implicit formula of

$$x^2y^2 + y^2z^2 + z^2x^2 - r^2xyz = 0.$$

Also, taking a parametrization of the sphere in terms of longitude (θ) and latitude (φ), gives parametric equations for the Roman surface as follows:

$$x = r^2 \cos \theta \cos \varphi \sin \varphi$$
$$y = r^2 \sin \theta \cos \varphi \sin \varphi$$
$$z = r^2 \cos \theta \sin \theta \cos^2 \varphi.$$

The origin is a triple point, and each of the *xy*-, *yz*-, and *xz*-planes are tangential to the surface there. The other places of self-intersection are double points, defining segments along each coordinate axis which terminate in six pinch points.

The entire surface has tetrahedral symmetry. It is a particular type (called type 1) of Steiner surface, that is, a 3-dimensional linear projection of the Veronese surface.

Derivation of Parametric Equations

Let a sphere have radius r, longitude φ, and latitude θ. Then its parametric equations are

$$x = r\cos\theta\cos\phi,$$
$$y = r\cos\theta\sin\phi,$$
$$z = r\sin\theta.$$

Then, applying transformation T to all the points on this sphere yields

$$x' = yz = r^2\cos\theta\sin\theta\sin\phi,$$
$$y' = zx = r^2\cos\theta\sin\theta\cos\phi,$$
$$z' = xy = r^2\cos^2\theta\cos\phi\sin\phi,$$

which are the points on the Roman surface. Let φ range from 0 to 2π, and let θ range from 0 to $\pi/2$.

Relation to the Real Projective Plane

The sphere, before being transformed, is not homeomorphic to the real projective plane, RP^2.

But the sphere centred at the origin has this property, that if point *(x,y,z)* belongs to the sphere, then so does the antipodal point *(-x,-y,-z)* and these two points are different: they lie on opposite sides of the centre of the sphere. The transformation T converts both of these antipodal points into the same point,

$$T:(x,y,z)\rightarrow(yz,zx,xy),$$

$$T:(-x,-y,-z)\rightarrow((-y)(-z),(-z)(-x),(-x)(-y))=(yz,zx,xy).$$

If this were true for only one or small subset of points of the sphere, then these points would just be double points. But since it is true of all points, then it is possible to consider the Roman surface to be homeomorphic to a "sphere modulo antipodes". However, these are not the only identifications that occur under this map. Consequently, the Roman surface is a quotient of the real projective plane $RP^2 = S^2 / (x\sim -x)$. Furthermore, this quotient has the special property that it is locally injective, making this an immersion of RP^2 into 3-space. It was previously stated that the Roman surface is homeomorphic to RP^2, but this was in error.

One-Sidedness

The Roman surface is non-orientable, i.e. one-sided. This is not quite obvious. To see this, look again at Figure.

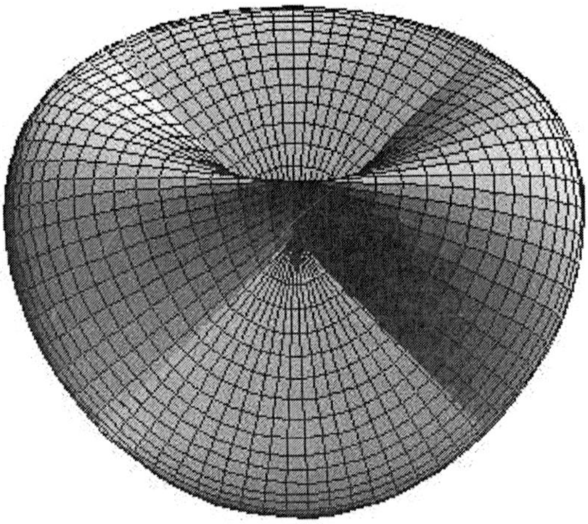

Imagine an ant on top of the "third" hyperbolic paraboloid, $z = x\ y$. Let this ant move North. As it moves, it will pass through the other two paraboloids, like a ghost passing through a wall. These other paraboloids only seem like obstacles due to the self-intersecting nature of the immersion. Let the ant ignore all double and triple points and pass right through them. So the ant moves to the North and falls off the edge of the world, so to speak. It now finds itself on the northern lobe, hidden underneath the third paraboloid. The ant is standing upside-down, on the "outside" of the Roman surface.

Let the ant move towards the Southwest. It will climb a slope (upside-down) until it finds itself "inside" the Western lobe. Now let the ant move in a Southeastern direction along the inside of the Western lobe towards the $z = 0$ axis, always above the x-y plane. As soon as it passes through the $z = 0$ axis the ant will be on the "outside" of the Eastern lobe, standing rightside-up.

Then let it move Northwards, over "the hill", then towards the Northwest so that it starts sliding down towards the $x = 0$ axis. As soon as the ant crosses this axis it will find itself "inside" the Northern lobe, standing right side up. Now let the ant walk towards the North. It will climb up the wall, then along the "roof" of the Northern lobe. The ant is back on the third hyperbolic paraboloid, but this time under it and standing upside-down. (Compare with Klein bottle.)

Double, Triple, and Pinching Points

The Roman surface has four "lobes". The boundaries of each lobe are a set of three lines of double points. Between each pair of lobes there is a line of double points. The surface has a total of three lines of double points, which lie (in the parametrization given earlier) on the coordinate axes. The three lines of double points intersect at a triple point which lies on the origin.

The triple point cuts the lines of double points into a pair of half-lines, and each half-line lies between a pair of lobes. One might expect from the preceding statements that there could be up to eight lobes, one in each octant of space which has been divided by the coordinate planes. But the lobes occupy alternating octants: four octants are empty and four are occupied by lobes.

If the Roman surface were to be inscribed inside the tetrahedron with least possible volume, one would find that each edge of the tetrahedron is tangent to the Roman surface at a point, and that each of these six points happens to be a *Whitney singularity*. These singularities, or pinching points, all lie at the edges of the three lines of double points, and they are defined by this property: that there is no plane tangent to surface at the singularity.

Alexander Horned Sphere

The Alexander horned sphere is a wild embedding of a sphere into space, discovered by J. W. Alexander (1924). It is the particular embedding of a sphere in 3-dimensional Euclidean space obtained by the following construction, starting with a standard torus:

1. Remove a radial slice of the torus.
2. Connect a standard punctured torus to each side of the cut, interlinked with the torus on the other side.
3. Repeat steps 1–2 on the two tori just added *ad infinitum*.

By considering only the points of the tori that are not removed at some stage, an embedding results of the sphere with a Cantor set removed. This embedding extends to the whole sphere, since points approaching two different points of the Cantor set will be at least a fixed distance apart in the construction.

The horned sphere, together with its inside, is a topological 3-ball, the Alexander horned ball, and so is simply connected; i.e., every loop can be shrunk to a point while staying inside. The exterior is *not* simply connected, unlike the exterior of the usual round sphere; a loop linking a torus in the above construction cannot be shrunk to a point without touching the horned sphere.

This shows that the Jordan–Schönflies theorem does not hold in three dimensions, as Alexander had originally thought. Alexander also proved that the theorem does hold in three dimensions for piecewise linear/smooth embeddings. This is one of the earliest examples where the need for distinction between the topological category of manifolds, and the categories of differentiable manifolds, and piecewise linear manifolds was noticed.

Now consider Alexander's horned sphere as an embedding into the 3-sphere, considered as the one-point compactification of the 3-dimensional Euclidean space R^3. The closure of the non-simply connected domain is called the solid Alexander horned sphere. Although the solid horned sphere is not a manifold, R. H. Bing showed that its double (which is the 3-manifold obtained by gluing two copies of the horned sphere together along the corresponding points of their boundaries) is in fact the 3-sphere.

One can consider other gluings of the solid horned sphere to a copy of itself, arising from different homeomorphisms of the boundary sphere to itself. This has also been shown to be the 3-sphere. The solid Alexander horned sphere is an example of a crumpled cube; i.e., a closed complementary domain of the embedding of a 2-sphere into the 3-sphere.

One can generalize Alexander's construction to generate other horned spheres by increasing the number of horns at each stage of Alexander's construction or considering the analogous construction in higher dimensions.

Other substantially different constructions exist for constructing such "wild" spheres. Another example, also found by Alexander, is Antoine's horned sphere, which is based on Antoine's necklace, a pathological embedding of the Cantor set into the 3-sphere.

Mapping Class Group

In mathematics, in the sub-field of geometric topology, the mapping class group is an important algebraic invariant of a topological space. Briefly, the mapping class group is a discrete group of 'symmetries' of the space.

Motivation

Consider a topological space, that is, a space with some notion of closeness between points in the space. We can consider the set of homeomorphisms from the space into itself, that is, continuous functions with continuous inverses: functions which stretch and deform the space continuously without puncturing or breaking the space. This set of homeomorphisms can be thought of as a space itself.

It can be seen fairly easily that this space forms a group under functional composition. We can also define a topology on this new space of homeomorphisms. The open sets of this new function space will be made up of sets of functions that map compact subsets K into open subsets U as K and U range throughout our original topological space, completed with their finite intersections (which must be open by definition of topology) and arbitrary unions (again which must be open).

This gives a notion of continuity on the space of functions, so that we can consider continuous deformation of the homeomorphisms themselves: called homotopies. We define the mapping class group by taking homotopy classes of homeomorphisms, and inducing the group structure from the functional composition group structure already present on the space of automorphisms.

Definition

The term mapping class group has a flexible usage. Most often it is used in the context of a manifold *M*. The mapping class group of *M* is interpreted as the group of isotopy-classes of automorphisms of *M*. So if *M* is a topological manifold, the mapping class group is the group of isotopy-classes of homeomorphisms of *M*. If *M* is a smooth manifold, the mapping class group is the group of isotopy-classes of diffeomorphisms of *M*. Whenever the group of automorphisms of an object *X* has a natural topology, the mapping class group of *X* is defined as $\mathrm{Aut}(X)/\mathrm{Aut}_0(X)$ where $\mathrm{Aut}_0(X)$ is the path-component of the identity in Aut(*X*).

For topological spaces, this is usually the compact-open topology. In the low-dimensional topology literature, the mapping class group of *X* is usually denoted MCG(*X*), although it is also frequently denoted $\pi_0(\mathrm{Aut}(X))$ where one substitutes for *Aut* the appropriate group for the category *X* is an object of. π_0 denotes the *0*-th homotopy group of a space.

So in general, there is a short-exact sequence of groups:

$$1 \to \mathrm{Aut}_0(X) \to \mathrm{Aut}(X) \to \mathrm{MCG}(X) \to 1.$$

Frequently this sequence is not split. If working in the homotopy category, the mapping-class group of *X* is the group of homotopy-classes of homotopy-equivalences of *X*.

There are many subgroups of mapping class groups that are frequently studied. If *M* is an oriented manifold, Aut(*M*) would be the orientation-preserving automorphisms of M and so the mapping class group of *M* (as an oriented manifold) would be index two in the mapping class group of M (as an unoriented manifold) provided M admits an orientation-reversing automorphism. Similarly, the subgroup that acts trivially on the homology of M is called the Torelli group of M, one could think of this as the mapping class group of a homologically-marked surface.

Mapping-Class Groups of Pairs

Given a pair of spaces (X,A) the mapping class group of the pair is the isotopy-classes of automorphisms of the pair, where an automorphism of (X,A) is defined as an automorphism of X that preserves A, i.e. $f: X \to X$ is invertible and $f(A) = A$.

Symmetry Group of Knot and Links

If K ,– S^3 is a knot or a link, the symmetry group of the knot (resp. link) is defined to be the mapping class group of the pair (S^3, K). The symmetry group of a hyperbolic knot is known to be dihedra or cyclic, moreover every dihedral and cyclic group can be realised as symmetry groups of knots. The symmetry group of a torus knot is known to be of order two Z_2.

Torelli Group

Notice that there is an induced action of the mapping class group on the homology (and cohomology) of the space X. This is because (co)homology is functorial and Homeo_0 acts trivially (because all elements are isotopic, hence homotopic to the identity, which acts trivially, and action on (co)homology is invariant under homotopy). The kernel of this action is the *Torelli group*. In the case of orientable surfaces, this is the action on first cohomology $H^1(\Sigma)$ E– Z^{2g}. Orientation-preserving maps are precisely those that act trivially on top cohomology $H^2(\Sigma)$ E– Z. $H^1(\Sigma)$ has a symplectic structure, coming from the cup product; since these maps are automorphisms, and maps preserve the cup product, the mapping class group acts as symplectic automorphisms, and indeed all symplectic automorphisms are realised, yielding the short exact sequence:

$$1 \to \text{Tor}(\Sigma) \to \text{MCG}(\Sigma) \to \text{Sp}(H^1(\Sigma)) \cong \text{Sp}_{2g}(\mathbf{Z}) \to 1$$

One can extend this to

$$1 \to \text{Tor}(\Sigma) \to \text{MCG}^*(\Sigma) \to \text{Sp}^{\pm}(H^1(\Sigma)) \cong \text{Sp}^{\pm}_{2g}(\mathbf{Z}) \to 1$$

The symplectic group is well-understood. Hence understanding the algebraic structure of the mapping class group often reduces to questions about the Torelli group.

Note that for the torus (genus 1) the map to the symplectic group is an isomorphism, and the Torelli group vanishes.

Stable Mapping Class Group

One can embed the surface $\Sigma_{g,1}$ of genus g and 1 boundary component into $\Sigma_{g+1,1}$ by attaching an additional hole on the end (i.e., gluing together and), and thus automorphisms of the small surface fixing the boundary extend to the larger surface. Taking the direct limit of these groups and inclusions yields the stable mapping class group, whose rational cohomology ring was conjectured by David Mumford (one of conjectures called the Mumford conjectures). The integral (not just rational) cohomology ring was computed in 2002 by Madsen and Weiss, proving Mumford's conjecture.

Dehn Twist

In geometric topology, a branch of mathematics, a Dehn twist is a certain type of self-homeomorphism of a surface (two-dimensional manifold).

Definition

Suppose that c is a simple closed curve in a closed, orientable surface S. Let A be a tubular neighbourhood of c. Then A is an annulus and so is homeomorphic to the Cartesian product of

$$S^1 \times I,$$

where I is the unit interval. Give A coordinates (s, t) where s is a complex number of the form

$$e^{i\theta}$$

with

$$\theta \in [0, 2\pi],$$

and t in the unit interval. Let f be the map from S to itself which is the identity outside of A and inside A we have

$$f(s,t) = (se^{i2\pi t}, t).$$

Then f is a Dehn twist about the curve c. Dehn twists can also be defined on a non-orientable surface S, provided one starts with a 2-sided simple closed curve c on S.

Mapping Class Group

It is a theorem of Max Dehn that maps of this form generate the mapping class group of isotopy classes of orientation-preserving homeomorphisms of any closed, oriented genus-g surface. W. B. R. Lickorish later rediscovered this result with a simpler proof and in addition showed that Dehn twists along $3g-1$ explicit curves generate the mapping class group (this is called by the punning name "Lickorish twist theorem"); this number was later improved by Stephen P. Humphries to $2g+1$, for $g>1$, which he showed was the minimal number.

Lickorish also obtained an analogous result for non-orientable surfaces, which require not only Dehn twists, but also "Y-homeomorphisms."

Nielsen–Thurston Classification

In mathematics, Thurston's classification theorem characterises homeomorphisms of a compact orientable surface. William Thurston's theorem completes the work initiated by Jakob Nielsen (1944).

Given a homeomorphism $f: S \to S$, there is a map g isotopic to f such that at least one of the following holds:

- g is periodic;
- g preserves some finite union of disjoint simple closed curves on S (in this case, g is called *reducible*); or
- g is pseudo-Anosov.

The case where S is a torus (i.e., a surface whose genus is one) is handled separately and was known before Thurston's work. If the genus of S is two or greater, then S is naturally hyperbolic, and the tools of Teichmüller theory become useful.

In what follows, we assume S has genus at least two, as this is the case Thurston considered. (Note, however, that the cases where S has boundary or is not orientable are definitely still of interest.)

The three types in this classification are not mutually exclusive, though a *pseudo-Anosov* homeomorphism is never *periodic* or *reducible*. A *reducible* homeomorphism g can be further analysed by cutting the surface along the preserved union of simple closed curves Γ.

Each of the resulting compact surfaces *with boundary* is acted upon by some power (i.e. iterated composition) of g, and the classification can again be applied to this homeomorphism.

The Mapping Class Group for Surfaces of Higher Genus

Thurston's classification applies to homeomorphisms of orientable surfaces of genus e− 2, but the type of a homeomorphism only depends on its associated element of the mapping class group *Mod(S)*. In fact, the proof of the classification theorem leads to a canonical representative of each mapping class with good geometric properties. For example:

- When g is periodic, there is an element of its mapping class that is an isometry of a hyperbolic structure on S.
- When g is pseudo-Anosov, there is an element of its mapping class that preserves a pair of transverse singular foliations of S, stretching the leaves of one (the *stable* foliation) while contracting the leaves of the other (the *unstable* foliation).

Mapping Tori

Thurston's original motivation for developing this classification was to find geometric structures on *mapping tori* of the type predicted by the Geometrization conjecture. The mapping torus M_g of a homeomorphism g of a surface S is the 3-manifold obtained from $S \times [0,1]$ by gluing $S \times \{0\}$ to $S \times \{1\}$ using g. The geometric structure of M_g is related to the type of g in the classification as follows:

- If g is periodic, then M_g has an $H^2 \times R$ structure;
- If g is reducible, then M_g has incompressible tori, and should be cut along these tori to yield pieces that each have geometric structures (the JSJ decomposition);
- If g is pseudo-Anosov, then M_g has a hyperbolic (i.e. H^3) structure.

The first two cases are comparatively easy, while the existence of a hyperbolic structure on the mapping torus of a pseudo-Anosov homeomorphism is a deep and difficult theorem (also due to Thurston).

The hyperbolic 3-manifolds that arise in this way are called *fibered* because they are surface bundles over the circle, and these manifolds are treated separately in the proof of Thurston's geometrization theorem for Haken manifolds. Fibered hyperbolic 3-manifolds have a number of interesting and pathological properties; for example, Cannon and Thurston showed that the surface subgroup of the arising Kleinian group has limit set which is a sphere-filling curve.

Fixed Point Classification

The three types of surface homeomorphisms are also related to the dynamics of the mapping class group *Mod(S)* on the Teichmüller space *T(S)*. Thurston introduced a compactification of *T(S)* that is homeomorphic to a closed ball, and to which the action of *Mod(S)* extends naturally. The type of an element *g* of the mapping class group in the Thurston classification is related to its fixed points when acting on the compactification of *T(S)*:

- If *g* is periodic, then there is a fixed point within *T(S)*; this point corresponds to a hyperbolic structure on *S* whose isometry group contains an element isotopic to *g*;
- If *g* is pseudo-Anosov, then *g* has no fixed points in *T(S)* but has a pair of fixed points on the Thurston boundary; these fixed points correspond to the *stable* and *unstable* foliations of *S* preserved by *g*.
- For some reducible mapping classes *g*, there is a single fixed point on the Thurston boundary; an example is a multi-twist along a pants decomposition *Γ*. In this case the fixed point of *g* on the Thurston boundary corresponds to *Γ*.

This is reminiscent of the classification of hyperbolic isometries into *elliptic*, *parabolic*, and *hyperbolic* types (which have fixed point structures similar to the *periodic*, *reducible*, and *pseudo-Anosov* types listed above).

Thurston Elliptization Conjecture

William Thurston's elliptization conjecture states that a closed 3-manifold with finite fundamental group is spherical, i.e. has a Riemannian metric of constant positive sectional curvature. A 3-manifold with such a metric is covered by the 3-sphere, moreover the group of covering transformations are isometries of the 3-sphere. Note that this means that if the original 3-manifold had in fact a trivial fundamental group, then it is homeomorphic to the 3-sphere (via the covering map).

Thus, proving the elliptization conjecture would prove the Poincaré conjecture as a corollary. In fact, the elliptization conjecture is logically equivalent to two simpler conjectures: the Poincaré conjecture and the spherical space form conjecture. The elliptization conjecture is a special case of Thurston's geometrization conjecture, which was proved in 2003 by G. Perelman.

Bibliography

Athanase Papadopoulos, *Metric Spaces, Convexity and Nonpositive Curvature*, European Mathematical Society, 2014.

Bourbaki, Nicolas: *Elements of Mathematics: General Topology*, Addison-Wesley, NY, 1996.

Burris, Stanley N., and H.P. Sankappanavar, H. P.: *A Course in Universal Algebra,* Springer-Verlag, 2001.

Conway, John B.: *A Course in Functional Analysis*, Springer-Verlag, 2014.

Dennis Zill, G.; Michael Cullen R., *Advanced engineering mathematics*, Jones & Bartlett Learning, 2019.

Dmitri Burago, Yu D Burago, Sergei Ivanov: *A Course in Metric Geometry*, American Mathematical Society, 2016.

Edward Kasner and James Newman: *Mathematics and the Imagination,* Simon and Schuster, New York, 1990.

Farkas, Hershel M.; Irwin Kra: *Riemann Surfaces*, Springer, New York, 2000.

George Arfken B. and Hans Weber, J., *Mathematical Methods for Physicists*, Academic Press, 2015.

Girault V. and Raviart, P.A., *Finite Element Methods for Navier–Stokes Equations: Theory and Algorithms,* Springer-Verlag, 2006.

Hatcher, A.: *"Algebraic Topology"*, Cambridge University Press, Cambridge, 2017.

Henri Cartan, Samuel Eilenberg: *Homological algebra*, Princeton University Press, Princeton, 2009.

Jacques Willy Delleur, *The handbook of groundwater engineering,* CRC Press, 2016.

Jeffreys, H. and Jeffreys, BS., *Methods of mathematical physics*, Cambridge University Press, 2009.

John Hopcroft E. and Jeffrey Ullman, D., *Introduction to Automata Theory, Languages and Computation*, Addison-Wesley Publishing, Reading Massachusetts, 2019.

Kelley, John L.: *General Topology*. Springer-Verlag, New York, 2005.

Lounesto, Pertti, *Clifford algebras and spinors*, Cambridge University Press, Cambridge, 2011.

Lynn Arthur Steen; J. Arthur Seebach jr: *Counterexamples in Topology. Berlin,* Springer-Verlag, New York, 2018.

Maunder, C. R. F.: *Algebraic Topology*, Van Nostrand Reinhold, London, 2010.

Megginson, Robert E.: *An Introduction to Banach Space Theory*, Springer, New York, 2018.

Michael Crowe J., *A History of Vector Analysis : The Evolution of the Idea of a Vectorial System*, Dover Publications, 2007.

Nahin, Paul J., *Oliver Heaviside: the life, work, and times of an electrical genius of the Victorian age*, JHU Press, 2016.

Peter Hilton; Stammbach, U. *A.: Course in homological algebra*, Springer-Verlag, New York, 2017.

Pickover, Clifford A.: *The Möbius Strip: Dr. August Möbius's Marvelous Band in Mathematics, Games, Literature, Art, Technology, and Cosmology*. Thunder's Mouth Press, 2016.

Rysxard Engelking: *General Topology*, Heldermann Verlag, Sigma Series in Pure Mathematics, 2009.

Schechter, Eric: *Handbook of Analysis and its Foundations*, Academic Press, San Diego, 2017.

Schey , H. M., *Div Grad Curl and all that: An informal text on vector calculus*, W. W. Norton & Company, 2015.

Stephen Willard: *General Topology*, Addison-Wesley Publishing Company, 2000.

Suthan Suthersan S., *Remediation engineering: design concepts*, CRC Press, 2016.

Ueno, Kenji et al.: *A Mathematical Gift: The Interplay Between Topology, Functions, Geometry, and Algebra*, American Mathematical Society, 2015.

Victor Bryant, *Metric Spaces: Iteration and Application*, Cambridge University Press, 2005.

Weisstein, Eric W., *CRC concise encyclopedia of mathematics*, CRC Press, 2013.

Whelan, P.M., Hodgeson, M.J., *Essential Principles of Physics,* John Murray, 2008.

Index

❑❑❑